A. F. Scheuerle · E. Schmidt
Atlas of Laser Scanning Ophthalmoscopy

Springer

Berlin
Heidelberg
New York
Hong Kong
London
Milan
Paris
Tokyo

A. F. Scheuerle · E. Schmidt

Atlas
of Laser Scanning
Ophthalmoscopy

Forewords by
H. E. Völcker · L. E. Pillunat · F. E. Kruse

With 22 Figures and 95 Plates,
Mostly in Color

Springer

ALEXANDER F. SCHEUERLE, MD
Universitäts-Augenklinik
Im Neuenheimer Feld 400
69120 Heidelberg
Germany

ECKART SCHMIDT, MD
Universitäts-Augenklinik
Fetscherstraße 4
01307 Dresden
Germany

ISBN 3-540-01868-9 Springer-Verlag Berlin Heidelberg New York

Library of Congress Cataloging-in-Publication Data
Scheuerle, A. F. (Alexander Friedrich)
Atlas of laser scanning ophthalmoscopy / A.F. Scheuerle, E. Schmidt.
p. ; cm.
Includes bibliographical references
ISBN 3-540-01868-9 (hardcover : alk. paper)
1. Scanning laser ophthalmoscopy–Atlases. 2. Confocal microscopy–Atlases. 3. Eye–Diseases–
Diagnosis–Atlases. 4. Glaucoma–Diagnosis–Atlases. I. Schmidt, E. (Eckart), 1970 II. Title.
[DNLM: 1. Microscopy, Confocal–methods–Atlases. 2. Ophthalmoscopy–methods–Atlases.
3. Glaucoma–diagnosis–Atlases. 4. Optic Disk–pathology–Atlases. 5. Optic Nerve Diseases–
diagnosis–Atlases. WW 17 S328a 2004]
RE79.S28S276 2004   617.7'1545–dc22   2003059100

Springer-Verlag is a part of Springer Science+Business Media

springeronline.com

© Springer-Verlag Berlin Heidelberg 2004
Printed in Germany

Cover-Design: e STUDIO CALAMAR, Pau/Girona, Spain
Typesetting and reproduction of the figures: AM-productions GmbH, Wiesloch, Germany
Printing and bookbinding: Stürtz AG, Würzburg, Germany

Printed on acid-free paper   24/3150PF   5 4 3 2 1 0

Foreword

With the advent of laser scanning ophthalmoscopy and the possibility of stereometric analysis of intraocular structures, ophthalmoscopy has entered a new dimension.

Herrmann von Helmholtz created the first ophthalmoscope in 1850 in Königsberg, East Prussia and taught at the University of Heidelberg between 1858 and 1870 as Professor of Physiology and Anatomy. In 1861, the diagnostic constellation of elevated intraocular pressure, papillary excavation, and visual field defects was defined as glaucoma. Shortly thereafter, Alfred Vogt showed that one could identify the nerve fiber layer of the retina with the ophthalmoscope.

The importance of documenting papillary and nerve fiber layer changes in glaucoma or suspected glaucoma, and the comparison of such changes over time, was rapidly recognized as being immensely important for this disease. The objective documentation of these anatomical structures was initially performed with photographic techniques, including papillary stereo photography and nerve fiber layer photography with red-free light.

Initial laboratory models of the laser scanning system for documentation of the ocular fundus were constructed in 1980 in the laboratories of R.H. Webb in Boston and J.F. Bille in Heidelberg. The close links between the University of Heidelberg and its industrial partners in the area of opto-electronics and computer technology allowed for the development of laser scanning ophthalmoscopy as a non-invasive means of documentation. Collaboration between the company Heidelberg Engineering and the University of Heidelberg's Department of Ophthalmology, with fundamental contributions by Professors R. Burk, F. Kruse, and K. Rohrschneider, allowed for the development of concepts which culminated in the valuable worldwide clinical use of the Heidelberg Retina Tomograph (HRT). The stereometric analysis of structures of the ocular fundus, especially the papilla, made possible by this device has led to its elemental use in documentation of the course of glaucomatous disease.

A. F. Scheuerle and E. Schmidt have not only contributed to further technical refinements of the second HRT generation but have also created the "Atlas of Laser Scanning Ophthalmoscopy" with the aid of the substantial experience and archives of the University of Heidelberg's Department of Ophthalmology. After a helpful introduction to the technical background and stereometric parameters of the Heidelberg Retina Tomograph, variations of the normal papilla are discussed and characterized. Understandably, glaucoma-related changes to the papilla represent the largest section of this book, followed by important information on longitudinal analysis, swelling of the optic disc, and stereometric changes of the retina. Throughout this work, topographic and intensity images are compared with conventional photography. Additionally, precise and didactic comments are easily identifiable through the use of a numeric system.

The "Atlas of Laser Scanning Ophthalmology" is an instructive and easy-to-use book, allowing for rapid aid in interpretation of HRT findings for every ophthalmologist. I wish the authors and publisher of this atlas well-earned success and wide circulation.

Heidelberg, September 2003

H. E. VÖLCKER, MD
Professor of Ophthalmology
and Chairman
Universitäts-Augenklinik
Heidelberg

Foreword

In recent decades it has been proven that morphological damage of the retinal nerve fiber layer and of the optic disc precedes psychophysically detectable glaucoma damage. Therefore, early glaucomatous damage might be detected much earlier by analysis of the retinal nerve fiber layer and the optic disc, whereas advanced glaucomatous damage might be better followed by automated perimetry. Besides early detection of glaucoma an exact follow-up of early glaucomatous optic disc changes is needed in order to adjust treatment in the case of progression.

Due to these considerations, optic nerve head analysis has been a major focus in glaucoma research for many years. Starting with morphometric methods based on optic disc photography, other techniques were evaluated. Stereovideoanalysis (ONH analyzer), pattern-assisted videography (glaucomascope) and other techniques did not prove sufficiently reliable. A new era of optic nerve head analysis started in 1988 with the technology of confocal laser scanning tomography (LTS).

Further refinements based on this technique led to the Heidelberg Retina Tomograph (HRT) models I and II. These devices have proven extremely sensitive in early glaucoma detection and are very reliable in glaucoma follow-up. The early prototypes like the LTS were hard to handle and measurements of the optic disc were very time consuming. Nowadays, both HRT devices can easily be used in daily practice and therefore represent valuable tools in glaucoma care. This volume contains chapters on the basic principles of the HRT, covers optic nerve head abnormalities and glaucomatous optic nerve changes, and focuses on follow-up examinations.

Therefore, the "Atlas of Laser Scanning Ophthalmoscopy" represents a suitable reference for all users of this technique as well as a valuable tool for all who are interested in this technology.

Dresden, September 2003

LUTZ E. PILLUNAT, MD
Professor of Ophthalmology
and Chairman
Universitäts-Augenklinik
Dresden

Foreword

More than 16 years ago the first commercially built confocal laser scanning instrument was introduced into ophthalmology by G. Zinser and U. Harbarth from the Heidelberg-based company Heidelberg Instruments. The prototype was archaic in terms of today's understanding of computer technology and required a physicist or at least an engineer for operation. Nevertheless, we were immediately fascinated by the enormous potential of laser scanning technology for the evaluation of the posterior pole of the eye and the optic nerve head. Following the publication of our first report on the very high reproducibility of measurements obtained by this technique, research in Heidelberg and elsewhere was focused on the development of evaluation strategies for optic nerve head measurements. Meanwhile the oldest database for confocal scanning laser investigations was established in Heidelberg originally using the old prototype instrument, the laser tomographic scanner LTS. For several years it was unclear whether this brilliant concept would be able to find its way into clinical routine ophthalmology primarily because of its prohibitive costs. It is clearly the merit of G. Zinser, the engineer behind confocal scanning laser instruments, to have refined the laser scanner into what is today's Heidelberg Retina Tomograph (HRT) and to have made the successful commercial realization possible.

Confocal laser scanning tomography has outgrown its infancy and is now used in hundreds of centers around the world. In contrast to the early days, the HRT devices have reached a level of reliability and a comfort of operation that makes it relatively easy for investigations to be performed. However, the interpretation of the results and even the evaluation of the quality of the obtained scanning series as basis for follow-ups are much more difficult. The need for a good atlas of laser scanning images is the result of the ever-increasing use of laser scanning technology. It is the logical consequence that A.F. Scheuerle and E. Schmidt have utilized the database that was initiated in Heidelberg and contains the files of several thousand patients in order to illustrate the entire spectrum of clinical findings that can be documented and quantified with the HRT I and II. Both authors have spent several years of their clinical training in the scanning laboratory and possess a great deal of expertise concerning educating ophthalmologists and their staff in training courses for HRT.

The "Atlas of Laser Scanning Ophthalmoscopy" aims to serve the needs of both the beginner and the expert user of the HRT. It contains both frequent and infrequent applications, and the authors have carefully selected a large quantity of images to illustrate a variety of clinical findings. We congratulate A. F. Scheuerle and E. Schmidt for their efforts in presenting the first atlas on laser scanning and providing the basis for the interpretation of laser scanning images that has been lacking in recent years.

Erlangen, September 2003

F. E. KRUSE, MD
Professor of Ophthalmology
and Chairman
Universitäts-Augenklinik
Erlangen

Contents

1   Introduction

A. F. Scheuerle, E. Schmidt

1.1   Why Is Laser Scanning Ophthalmoscopy Needed?

Glaucoma is one of the leading causes of blindness. The prevalence of glaucoma is between 1–2% and approximately 50% of glaucoma patients around the world remain undiagnosed. The clinical evaluation of glaucoma patients usually includes tonometry, perimetry and assessment of the optic nerve head. Elevated intraocular pressure (IOP) is widely recognized as the most important risk factor for the development of glaucomatous neuropathy. Recent studies show individual differences in tolerances towards certain levels and amplitudes of IOP. In most cases, mere IOP measurements are not sufficient, neither for the diagnosis nor for the monitoring of glaucomatous disease. Perimetry represents an excellent test of optic nerve function. There is a tremendous amount of reliable equipment that uses efficient strategies to detect and monitor visual field defects characteristic of glaucoma. However, perimetry is a psychophysical test associated with a relatively high variation of results, even in the case of good cooperation of the patient and, most important, patients with early stage glaucoma or suspected glaucoma almost never manifest visual field defects. Therefore, suprathreshold visual field testing is known to have a high specificity (90%) combined with a low sensitivity (52%) in detecting glaucoma cases (Katz et al. 1993). There is some evidence that damage to the optic disc and the retinal nerve fiber layer (RNFL) may precede the detection of abnormal visual fields (Sommer et al. 1991). In the hands of skilled ophthalmologists, biomicroscopic funduscopy offers a fast, qualitative and still indispensable evaluation of the optic disc. Especially for early diagnosis and long term follow-up, the *detection and quantification of disc changes* in the course of glaucoma is crucial. Confocal laser scanning ophthalmoscopy provides objective, three-dimensional and quantitative assessment of the living anatomy of the optic nerve head. Its unprecedented accuracy and reproducibility fit the diagnostic needs of modern glaucoma management.

1.2   Technical Background

The development of confocal laser scanning microscopy in the late 1980s enabled physicians to receive precise three-dimensional images of the optic nerve head in vivo (Zinser et al. 1988). Confocal laser scanning ophthalmoscopes soon became valuable tools for clinical use to obtain objective and quantitative data of the optic nerve head and the peripapillary retina. Images are obtained non-invasively, rapidly and with low-level illumina-

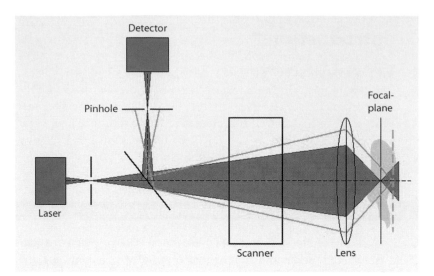

Fig. 1. Principle of confocality

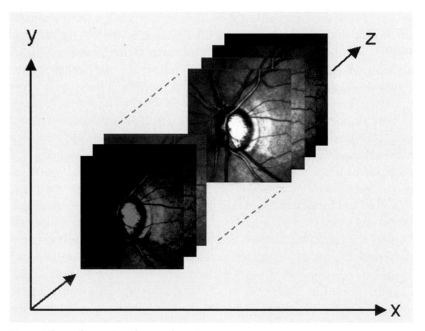

Fig. 2. Three-dimensional optical sections

tion. The laser source employed in the Heidelberg Retina Tomograph (HRT) is a diode laser with a wavelength of 670 nm. Usually there is no need to dilate the pupil of the eye under examination.

The precision of laser scanning ophthalmoscopy is based on the optical principle of confocality shown in Fig. 1. A single point of laser light is projected through a pinhole onto the focal plane of interest at the posterior pole of the eye. The laser light is reflected back through the second confocal pinhole and onto the light-sensitive detector. Any signal outside the focal plane will be blocked by the detector pinhole. To acquire a two-dimensional image (optical section) of the retina, the laser beam is periodically deflected by oscillating mirrors. Each of the two-dimensional images consists of 256×256 (HRT II 384×384) picture elements (pixels). If the position of the focal plane is moved to different depths along the optic nerve (z-axis) and further optical sections are acquired, the result will be a layered three-dimensional image (tomography) as demonstrated in Figs. 2 and 3. The HRT generates a series of 32 (HRT II 16–64) consecutive and equidistant two-dimensional optical section images Fig. 4. From the distribution of the amount of reflected light along the z-axis in the three-dimensional images, the height of the retinal surface is computed at each point. Such a topography consists of more than 65,000 (HRT II>147,000) independent local height measurements and provides imaging data with high spatial resolution (about 10 μm per pixel). If at least three topographies are used to compute a mean topography, the reproducibility of local height measurements at each measurement location in a mean topography image will be around 20 μm for healthy and glaucomatous eyes (Kruse et al. 1989, Weinreb et al. 1993, Janknecht et al. 1994, Dannheim et al. 1995). The coefficients of variation of the stereometric parameters are typically around 5 %. (Mikelberg et

Table 1. HRT and HRT II specifications

Type of scanner	HRT	HRT II
Transverse field of view	10°×10°, 15°×15° or 20°×20°	15°×15° (fixed)
Longitudinal field of view	0.5–4.0 mm	1.0–4.0 mm (automatic)
Digital image size 2D image 3D image	256×256 pixels 256×256×32 voxels	384×384 pixels 384×384×16 to 384×384×64 voxels
Acquisition time 2D image 3D image	0.032 s 1.4 s	0.025 s 1.0 s (2 mm depth)
Focus range	–12 to +12 diopters	–12 to +12 diopters
Optical resolution (limited by the eye) Transverse Longitudinal	10 μm 300 μm	10 μm 300 μm
Digital resolution Transverse Longitudinal	10–20 μm/pixel 16 μm–128 μm/plane	10 μm/pixel 62 μm/plane
Fixation aid	External	Internal
Correction of brightness and sensitivity	Manual	Automatic
Laser source	Diode laser, 675 nm	Diode laser, 675 nm
Software platform	DOS	Windows
Commercialized in	1992	1999
Price (US$)	ca. 60,000	ca. 35,000
Units sold (worldwide)	350/7 years	3,000/3.5 years

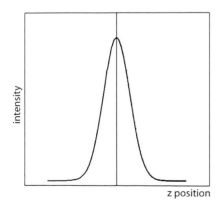

Fig. 3. Intensity of reflected light along the z-axis

al. 1993; Rohrschneider et al. 1994). Table 1 summarizes the technical characteristics of HRT and the current model HRT II.

In 1988, the Laser Tomographic Scanner (LTS), the predecessor of the HRT, was the first laser scanning ophthalmoscope capable of optical tomographies of the anterior and posterior parts of the eye. Secondary to its enormous size and price (ca. $300,000) only 5 units of the LTS were produced.

1.3 Stereometric Parameters of the HRT

The standard HRT II software displays 22 stereometric parameters. Table 2 presents an overview of HRT parameters of healthy optic discs and in different stages of glaucoma. As most of these parameters' names are self-explaining and all parameters are well described in the HRT handbook, only two will be further discussed in this chapter. The reference plane is proba-

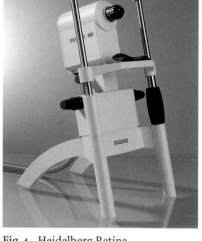

Fig. 4. Heidelberg Retina Tomograph II

Table 2. Typical values of HRT parameters of optic discs in the course of glaucoma (Burk et al. 2001)

Parameter	Normal (n=349)	Early (MD 2–5 dB) (n=192)	Moderate (MD 5–10 dB) (n=97)	Advanced (MD>10 dB) (n=105)
Disc area (mm^2)	2.257±0.563	2.346±0.569	2.310±0.554	2.261±0.416
Cup area (mm^2)	0.768±0.505	0.953±0.594	1.051±0.647	1.445±0.562
Rim area (mm^2)	1.489±0.291	1.393±0.340	1.260±0.415	0.817±0.334
Cup volume (mm^3)	0.240±0.245	0.294±0.270	0.334±0.318	0.543±0.425
Rim volume (mm^3)	0.362±0.124	0.323±0.156	0.262±0.139	0.128±0.096
Cup/disc area ratio	0.314±0.152	0.380±0.179	0.430±0.203	0.621±0.189
Horizontal cup/disc ratio	0.567±0.200	0.623±0.221	0.658±0.226	0.808±0.185
Vertical cup/disc ratio	0.460±0.206	0.538±0.214	0.573±0.226	0.756±0.194
Mean cup depth (mm)	0.262±0.118	0.279±0.115	0.289±0.130	0.366±0.182
Maximum cup depth (mm)	0.679±0.223	0.680±0.210	0.674±0.249	0.720±0.276
Cup shape measurement	−0.181±0.092	−0.147±0.098	−0.122±0.095	−0.036±0.096
Height variation contour (mm)	0.384±0.087	0.364±0.100	0.330±0.108	0.256±0.090
Mean RNFL thickness (mm)	0.244±0.063	0.217±0.076	0.182±0.086	0.130±0.061
RNFL cross-sectional area (mm^2)	1.282±0.328	1.155±0.396	0.957±0.440	0.679±0.302

bly the most important variable because it separates the cup and neuroretinal rim. Its position influences the majority of parameter values significantly. Whereas during the first years of laser scanning tomography a fixed offset reference plane (320 μm below the mean retina height) was commonly used, the so called "standard reference plane" (SRP) is now part of the regular HRT software. The SRP is defined by an optic disc border contour-line segment of 6° width (350°- 356°) corresponding to the site of the papillomacular bundle which is considered to remain relatively stable in the course of glaucoma. The reproducibility of the SRP-segment height measurements was 16.0±10.8 μm for normal eyes and 23.4±18.0 for glaucoma eyes. To ensure that the automatic reference level determination for intrapapillary parameters remained below the disc border height, the SRP level was defined at a 50 μm offset, representing >2 standard deviations of average segment height reproducibility in glaucoma. In contrast to a fixed offset reference plane, the SRP respects the interindividual variability of optic disc topography at the expense of the need for an accurate (optic disc border) contour-line (Burk et al. 2000). Figure 5 illustrates the position of the SRP.

One of the few features that are independent from a reference plane is the retinal surface height profile along the drawn contour line. The height profile demonstrated by the HRT software always starts temporal at 0°. The increased thickness of nerve fiber bundles at the site of the superior and inferior poles of a normal optic disc generate the characteristic "double hump" configuration of the retinal surface height profile (Burk et al. 1990). "Height variation contour" and "mean retinal nerve fiber layer thickness" are two quantitative parameters directly derived from the dynamics of the double hump profile. The "height variation contour" is calculated by the difference between the most elevated and most depressed point of the contour line and is therefore independent from the SRP. The "mean retinal nerve fiber layer thickness" corresponds to the mean height difference between the SRP and the height profile along the contour line. Furthermore, the position of the double hump contour-line with respect to the mean retina height (MRH) and the SRP can be used as a quantitative diagnostic criterion in the evaluation of HRT tomographies. In the current HRT II software, the position of the MRH is always defined as a zero point of the height (0.0 mm z-axis), marked by a horizontal black line. While in normal eyes the polar peaks usually reach the MRH, the contour line of the disc border

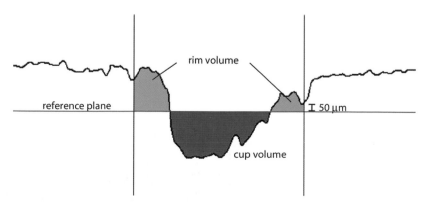

Fig. 5. Position of the standard reference plane

in glaucoma is typically depressed below the MRH. However, in cases with generalized depression of the retinal surface caused by global atrophy of the retinal nerve fiber layer and indicated by low SRP values ($<253\,\mu m$), one or two polar peaks may reach the MRH even if glaucomatous damage is present (Scheuerle et al. 2001).

2 Healthy Optic Discs

A. F. SCHEUERLE, E. SCHMIDT

2.1 Natural Variety of Optic Discs

The human optic disc presents a large natural variety of different sizes and shapes which can sometimes make the diagnosis of glaucoma more difficult. Considering that different machines like fundus cameras and laser scanning ophthalmoscopes exhibit statistically significant systematic differences in optic disc sizes of comparable populations (Meyer et al. 2001), we conclude that the proportions of stereometric disc parameters are much of more clinical interest than the validity of pure size measurements. Therefore, we will briefly comment on three relatively common types of optic discs that might confuse the HRT user because their anatomy does not fit the scheme which automated classification procedures are based on.

Megalopapillas are defined by large disc areas (>3.0 mm²) that are typically associated with large cup areas, normal visual fields and normal IOP (Sampaolesi R et al. 2001). Secondary to the unusual dimensions, even in healthy megalopapillas, the cup areas often appear enlarged and glaucomatous. In most cases, funduscopy shows a round cup shape, not characteristic for glaucoma. Laser scanning tomography allows precise analysis of the stereometric parameters and reveals that rim areas and rim volumes of healthy megalopapillas are similar to those of regular sized optic discs. The larger circumference of a megalodisc leads to a "horizontally stretched" contour line that, therefore, seems to be flatter compared with the contour line of a normal disc. Leaving brightness control to the automatic mode of HRT II, one might notice that topographies of megalodiscs tend to be underexposed due to the intense reflection of the large cup area. In such cases, the exact border of the optic disc is hardly visible. Correct placement of the contour line also becomes more difficult while the use of the interactive mode is encouraged. Horizontal (cross-section) height profiles sometimes show "steps" in the temporal area of the peripapillary retinal surface which may not be included by drawing the contour line (Figs. 6–9).

We recommend the three-dimensional reconstruction to validate any questionable contour line placement (Fig. 10). Despite normal rim volumes, automated classification procedures may rate megalodiscs as glaucomatous. In our experience, the Moorfields regression analysis often detects sectors outside normal limits in the nasal portion of megalopapillas. Having a normal global classification, these "pathologic" nasal sectors seem to be clinically irrelevant in the entity of megalopapillas. In megalopapillas, automated classification procedures show a higher sensitivity but a lower specificity compared to the published figures based on relatively small study populations.

On the contrary, endangered micropapillas ($<1.9\,mm^2$) are sometimes misjudged by clinicians because their already glaucomatous cup areas appear to be small. Laser scanning tomography and automated classisification procedures provide a more objective evaluation of the cup-disc-ratio and other stereometric parameters.

We noted that many micropapillas are also tilted discs. This is an unfortunate combination as the anatomy of a tilted disc makes the diagnosis of glaucoma even more difficult. The tilted disc belongs to the few forms of optic nerve heads that are not suitable for any currently used automated classification procedure. In a tilted disc, the temporal part of the disc border that is used to define the standard reference plane lies on a significantly lower level than the nasal part. The resulting reference plane is, therefore, only valid for the temporal section of the optic disc but is, of course, applied to calculate stereometric parameters concerning the whole disc area. The nasal portion than lies far above the reference plane simulating excessive rim area, rim volume, etc., which provokes false classification results (i.e., the classification tends to indicate a normal optic disc even in case of visible glaucomatous damage). The tilted surface produces vertically pronounced cup shapes, usually with a small and flat excavation. The contour lines of tilted discs demonstrate high amplitudes and sometimes asymmetry.

In summary, it can be said that laser scanning tomography does not offer great help classifying suspicious tilted discs. However, we want to emphasize that the advantages of laser scanning tomography remain unaffected regarding the benefit of objective long-term monitoring, especially in optic nerve heads that are difficult to evaluate due to their extraordinary anatomy. Even in the case of congenital optic nerve head anomalies, the analysis of sequential optic disc images will usually reveal the progress of glaucomatous disease.

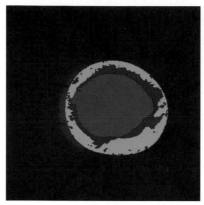

Fig. 6. Topography image of a megalo-papilla after correct contour line placement

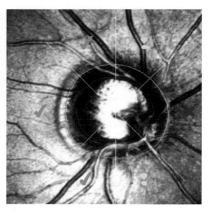

Fig. 7. Intensity image of a megalo-papilla after correct contour line placement

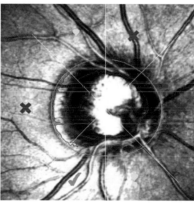

Fig. 8. Intensity image of the same megalopapilla after *incorrect* contour line placement including the temporal "step"

Fig. 9. Correspondig topography image. The extension of the temporal disc border provokes a glaucomatous configuration of the cup

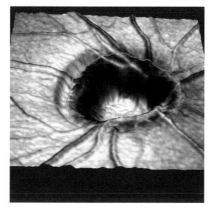

Fig. 10. Three-dimensional reconstruction demonstrating correct placement of the contour line along the borders of the megalopapilla

Plate 1

Comments

- Circular, relatively small optic disc (1)
- Increased global "rim area" and "rim volume"
- Small "cup/disc area ratio" with normal "linear cup/disc ratio"
- Small cup area with deep excavation (2)
- Normal "cup shape measure" value despite steep nasal cup border (3) due to central vessel trunk
- Moorfields analysis rates all sectors as "within normal limits"
- Vital appearance of the peripapillary retinal nerve fiber layer with excellent reflectivity (4)
- Dynamic and symmetric height profile of the contour line (5)
- Both polar peaks ("double humps") (6) reach mean retina height (*)

Original printout with:
a Topography image, OD
b Intensitiy image, OD

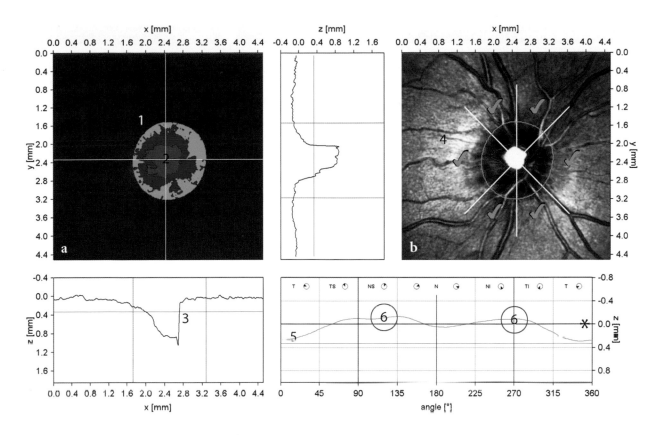

Stereometric Analysis ONH	
Disk Area	2.037 mm²
Cup Area	0.401 mm²
Rim Area	1.636 mm²
Cup Volume	0.117 cmm
Rim Volume	0.470 cmm
Cup/Disk Area Ratio	0.197
Linear Cup/Disk Ratio	0.444
Mean Cup Depth	0.232 mm
Maximum Cup Depth	0.889 mm
Cup Shape Measure	-0.342
Height Variation Contour	0.425 mm
Mean RNFL Thickness	0.308 mm
RNFL Cross Sectional Area	1.561 mm²
Reference Height	0.332 mm
Topography Std Dev.	13 μm

Plate 2

Comments

- Medium-sized optic disc (1)
- Normal global "rim area" and "rim volume" with wide neuroretinal rim (2)
- Normal "cup/disc area ratio" and slightly increased "linear cup/disc ratio"
- Medium-sized central cup with horizontally pronounced oval shape (3)
- Borderline "cup shape measure" value with steep inferior cup border (4) due to transverse vessels
- Normal "height variation contour" value
- Moorfields analysis rates all sectors as "within normal limits"
- Vital appearance of the peripapillary retinal nerve fiber layer with excellent reflectivity (5)
- Dynamic and symmetric height profile of the contour line (6)
- Both polar peaks ("double humps") (7) reach mean retina height (∗)

Original printout with:
a Topography image, OD
b Intensitiy image, OD

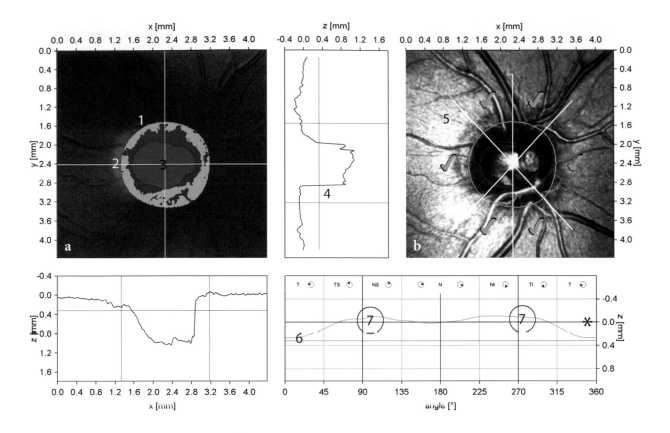

Stereometric Analysis ONH	
Disk Area	2.500 mm²
Cup Area	0.926 mm²
Rim Area	1.574 mm²
Cup Volume	0.436 cmm
Rim Volume	0.447 cmm
Cup/Disk Area Ratio	0.370
Linear Cup/Disk Ratio	0.609
Mean Cup Depth	0.454 mm
Maximum Cup Depth	1.025 mm
Cup Shape Measure	-0.123
Height Variation Contour	0.376 mm
Mean RNFL Thickness	0.296 mm
RNFL Cross Sectional Area	1.657 mm²
Reference Height	0.317 mm
Topography Std Dev.	12 μm

Comments

- Small, tilted optic disc (1) with prominent vessel trunk
- Increased global "rim area" and "rim volume" secondary to significant height differences of temporal and nasal disc borders (2)
- Very small "cup/disc area ratio" and small "linear cup/disc ratio"
- Very small eccentric cup with shallow excavation (3)
- Normal "cup shape measure" value
- Normal "height variation contour" value
- Low standard reference plane (red line)
- Vital appearance of the peripapillary retinal nerve fiber layer with excellent reflectivity (4)
- Dynamic and symmetric height profile of the contour line (5) significantly above mean retina height (*)

Original printout with:
a Topography image, OD
b Intensitiy image, OD
c Photograph of the optic disc, OD

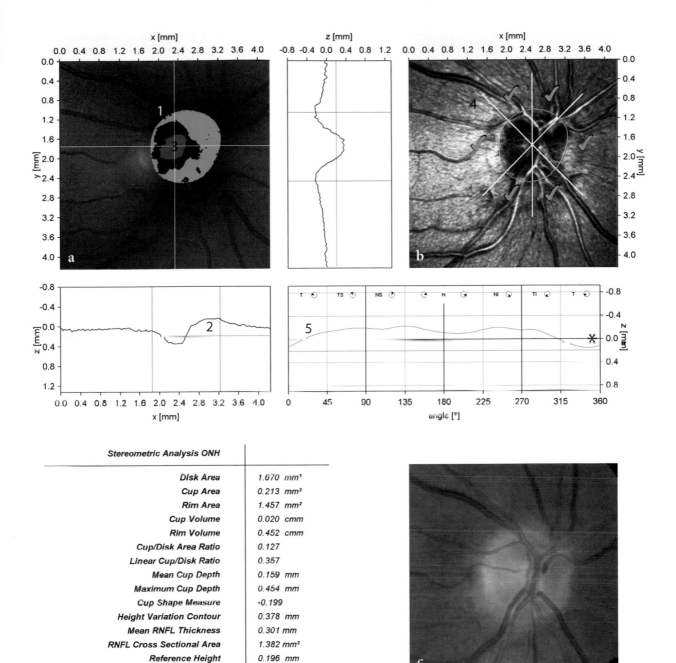

Stereometric Analysis ONH	
Disk Area	1.670 mm²
Cup Area	0.213 mm²
Rim Area	1.457 mm²
Cup Volume	0.020 cmm
Rim Volume	0.452 cmm
Cup/Disk Area Ratio	0.127
Linear Cup/Disk Ratio	0.357
Mean Cup Depth	0.159 mm
Maximum Cup Depth	0.454 mm
Cup Shape Measure	-0.199
Height Variation Contour	0.378 mm
Mean RNFL Thickness	0.301 mm
RNFL Cross Sectional Area	1.382 mm²
Reference Height	0.196 mm
Topography Std Dev.	9 µm

Plate 4

Comments

- Small oval shaped optic disc (1) with prominent vessel trunk (2)
- Increased global "rim area" and "rim volume" secondary to height differences of temporal and nasal disc borders (3)
- Very small "cup/disc area ratio" and "linear cup/disc ratio"
- Cup area is very small and barely visible with shallow excavation (4)
- Normal "cup shape measure" value
- Low standard reference plane (red line) with small distance to mean retina height (∗)
- Disc borders are difficult to define
- Vital appearance of the peripapillary retinal nerve fiber layer with excellent reflectivity (5)
- Dynamic and slightly asymmetric height profile of the contour line (6) far above mean retina height (∗)

Original printout with:
a Topography image, OS
b Intensitiy image, OS
c 3D-image, OS

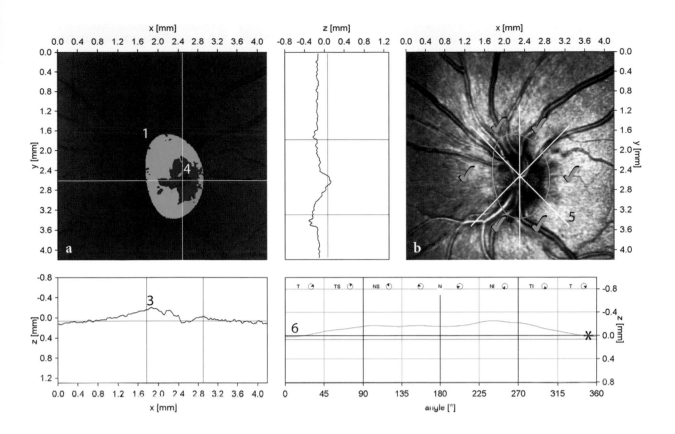

a

b

x [mm]
0.0 0.4 0.8 1.2 1.6 2.0 2.4 2.8 3.2 3.6 4.0

z [mm]
-0.8 -0.4 0.0 0.4 0.8 1.2

x [mm]
0.0 0.4 0.8 1.2 1.6 2.0 2.4 2.8 3.2 3.6 4.0

y [mm]

1
4

5

3

T | TS | NS | N | NI | TI | T

6

x [mm]
0.0 0.4 0.8 1.2 1.6 2.0 2.4 2.8 3.2 3.6 4.0

angle [°]
0 45 90 135 180 225 270 315 360

Stereometric Analysis ONH	
Disk Area	1.522 mm²
Cup Area	0.031 mm²
Rim Area	1.491 mm²
Cup Volume	0.001 cmm
Rim Volume	0.333 cmm
Cup/Disk Area Ratio	0.021
Linear Cup/Disk Ratio	0.143
Mean Cup Depth	0.055 mm
Maximum Cup Depth	0.191 mm
Cup Shape Measure	-0.253
Height Variation Contour	0.274 mm
Mean RNFL Thickness	0.197 mm
RNFL Cross Sectional Area	0.867 mm²
Reference Height	0.062 mm
Topography Std Dev.	15 µm

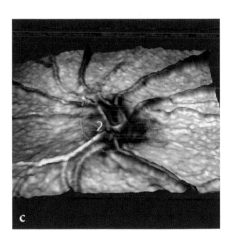

c

2

Comments

- Very large optic disc (1)
- Vessel trunk simulates increased neuroretinal rim in the nasal inferior sector (2)
- Temporal disc border lies exactly on the scleral ring (3)
- Increased global "rim area" and "rim volume"
- Increased "cup/disc area ratio" and "linear cup/disc ratio"
- Large cup with deep excavation (4)
- Pathologic "cup shape measure" value
- Increased "height variation contour" value
- Very high standard reference plane (red line)
- Moorfields analysis rates all sectors, except the temporal inferior one, as "within normal limits"
- Vital appearance of the peripapillary retinal nerve fiber layer with excellent reflectivity (5)
- Dynamic and slightly asymmetric height profile of the contour line (6)
- Both polar peaks ("double humps") (7) do not reach mean retina height (∗)

Original printout with:
a Topography image, OD
b Intensitiy image, OD
c 3D-image, OD

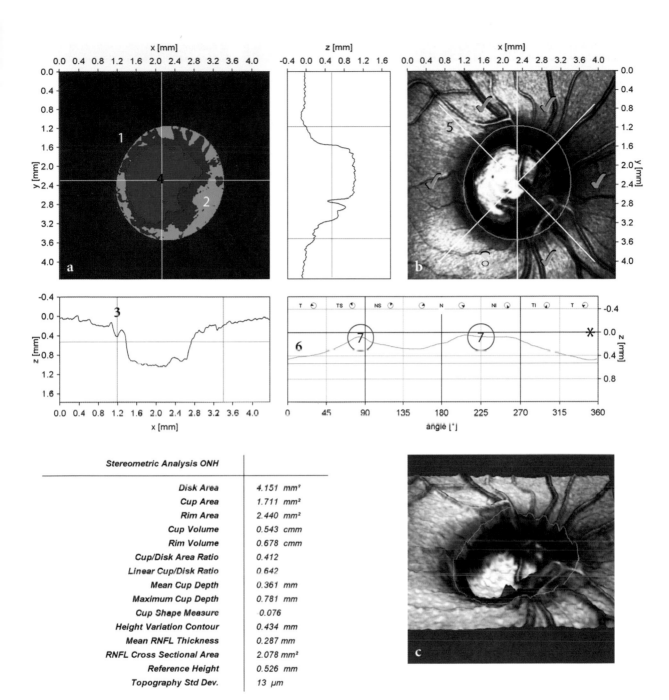

Stereometric Analysis ONH	
Disk Area	4.151 mm²
Cup Area	1.711 mm²
Rim Area	2.440 mm²
Cup Volume	0.543 cmm
Rim Volume	0.678 cmm
Cup/Disk Area Ratio	0.412
Linear Cup/Disk Ratio	0.642
Mean Cup Depth	0.361 mm
Maximum Cup Depth	0.781 mm
Cup Shape Measure	-0.076
Height Variation Contour	0.434 mm
Mean RNFL Thickness	0.287 mm
RNFL Cross Sectional Area	2.078 mm²
Reference Height	0.526 mm
Topography Std Dev.	13 μm

Plate 6

Comments

- Large optic disc (1)
- "Step" in the temporal peripapillary retina (2)
- Increased global "rim area" and "rim volume"
- Increased "cup/disc area ratio" and "linear cup/disc ratio"
- Large cup with deep central excavation and horizontally pronounced oval shape (3)
- Pathologic "cup shape measure" value
- Increased "height variation contour" value
- High standard reference plane (red line)
- Moorfields analysis rates the three nasal sectors as "out of normal limits"
- Vital appearance of the peripapillary retinal nerve fiber layer with excellent reflectivity (4)
- Dynamic and slightly asymmetric height profile of the contour line (5) just below the mean retina height (∗)
- Superior polar peak (6) does not reach mean retina height (∗)

Original printout with:
a Topography image, OD
b Intensitiy image, OD
c 3D-image, OD

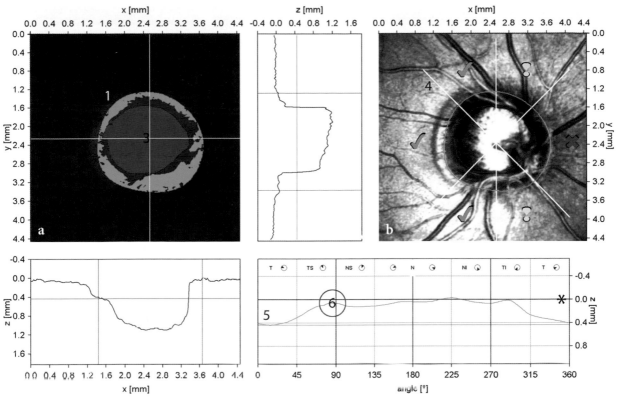

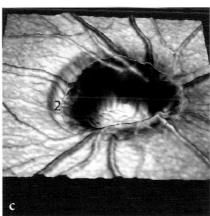

Stereometric Analysis ONH	
Disk Area	3.830 mm²
Cup Area	1.929 mm²
Rim Area	1.901 mm²
Cup Volume	0.918 cmm
Rim Volume	0.565 cmm
Cup/Disk Area Ratio	0.504
Linear Cup/Disk Ratio	0.710
Mean Cup Depth	0.542 mm
Maximum Cup Depth	1.048 mm
Cup Shape Measure	-0.028
Height Variation Contour	0.467 mm
Mean RNFL Thickness	0.284 mm
RNFL Cross Sectional Area	1.972 mm²
Reference Height	0.435 mm
Topography Std Dev.	11 µm

Comments

- Large optic disc (1)
- "Step" in the temporal peripapillary retina (2)
- Very large global "rim area" and large "rim volume"
- Increased "cup/disc area ratio" and "linear cup/disc ratio"
- Relatively small, slightly eccentric cup of round shape (3)
- Normal "cup shape measure" value
- Normal "height variation contour" value
- Normal standard reference plane height (red line)
- Moorfields analysis rates all sectors as "within normal limits"
- Vital appearance of the peripapillary retinal nerve fiber layer with excellent reflectivity (4)
- Dynamic and slightly asymmetric height profile of the contour line (5)
- Both polar peaks ("double humps") (6) reach mean retina height (∗)

Original printout with:
a Topography image, OD
b Intensitiy image, OD
c 3D-image, OD

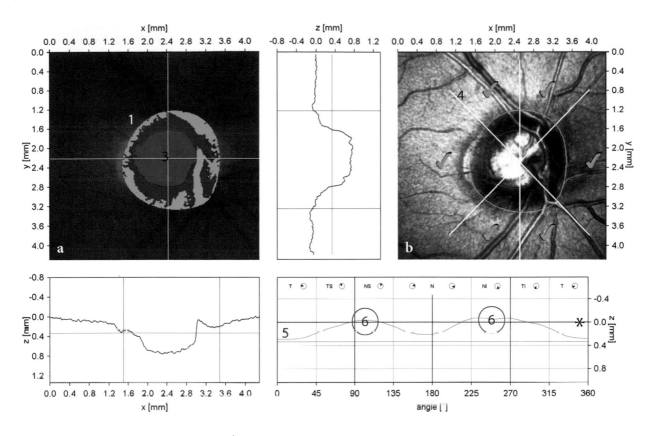

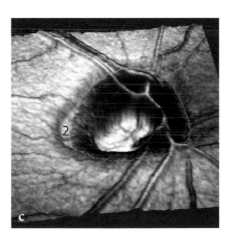

Stereometric Analysis ONH	
Disk Area	3.314 mm²
Cup Area	1.162 mm²
Rim Area	2.152 mm²
Cup Volume	0.328 cmm
Rim Volume	0.487 cmm
Cup/Disk Area Ratio	0.351
Linear Cup/Disk Ratio	0.592
Mean Cup Depth	0.278 mm
Maximum Cup Depth	0.684 mm
Cup Shape Measure	-0.154
Height Variation Contour	0.364 mm
Mean RNFL Thickness	0.238 mm
RNFL Cross Sectional Area	1.537 mm²
Reference Height	0.331 mm
Topography Std Dev.	13 μm

Plate 8

Comments

- Very large optic disc (1)
- "Step" in the temporal peripapillary retina (2)
- Very large global "rim area" and "rim volume"
- Increased "cup/disc area ratio" and "linear cup/disc ratio"
- Large cup with very deep central excavation of round shape (3)
- Pathologic "cup shape measure" value
- Increased "height variation contour" value
- Moorfields analysis rates all sectors, exept the temporal one, as "out of normal limits"
- Vital appearance of the peripapillary retinal nerve fiber layer with excellent reflectivity (4)
- Edema map demonstrates a circular intact neuroretinal rim (5)
- Highly dynamic and slightly asymmetric height profile of the contour line (6) just below mean retina height (✱)

a Intensitiy image, OS
b Edema map, OS
c Topography image, OS
d Profile of contour line
e Stereometric parameters

Plate 9

Comments

- Large optic disc (1)
- "Step" in the temporal peripapillary retina (2)
- Very large global "rim area" and "rim volume"
- Normal "cup/disc area ratio" and "linear cup/disc ratio"
- Relativly small cup with slightly eccentric excavation of round shape (3)
- Pathologic "cup shape measure" value
- Increased "height variation contour" value
- Moorfields analysis rates all sectors as "within normal limits"
- Vital appearance of the peripapillary retinal nerve fiber layer with excellent reflectivity (4)
- Dynamic and symmetric height profile of the contour line (5)
- Both polar peaks ("double humps") (6) reach mean retina height (✱)

a Intensitiy image, OD
b Photograph, OD
c Topography image, OD
d Profile of contour line
e Stereometric parameters

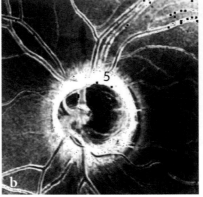

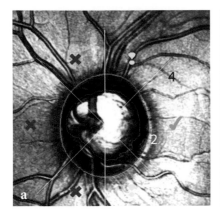

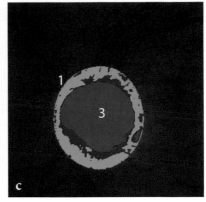

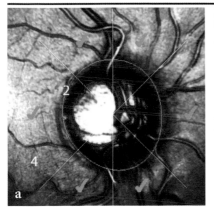

Parameters	global
disc area [mm²]	4.419
cup area [mm²]	2.328
rim area [mm²]	2.092
cup/disc area ratio [ ]	0.527
rim/disc area ratio [ ]	0.473
cup volume [mm³]	1.604
rim volume [mm³]	0.678
mean cup depth [mm]	0.744
maximum cup depth [mm]	1.308
height variation contour [mm]	0.491
cup shape measure [ ]	0.045
mean RNFL thickness [mm]	0.277
RNFL cross sectional area [mm²]	2.065
linear cup/disc ratio [ ]	0.726
maximum contour elevation [mm]	0.001
maximum contour depression [mm]	0.522
CLM temporal-superior [mm]	0.348
e CLM temporal-inferior [mm]	0.183

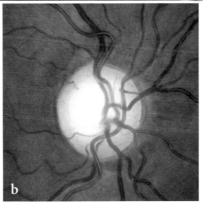

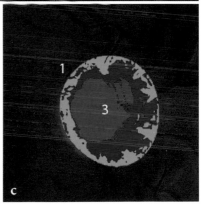

Parameters	global
disc area [mm²]	3.832
cup area [mm²]	1.308
rim area [mm²]	2.524
cup/disc area ratio [ ]	0.341
rim/disc area ratio [ ]	0.659
cup volume [mm³]	0.170
rim volume [mm³]	0.666
mean cup depth [mm]	0.247
maximum cup depth [mm]	0.541
height variation contour [mm]	0.434
cup shape measure [ ]	-0.079
mean RNFL thickness [mm]	0.304
RNFL cross sectional area [mm²]	2.110
linear cup/disc ratio [ ]	0.584
maximum contour elevation [mm]	-0.036
maximum contour depression [mm]	0.397
CLM temporal-superior [mm]	0.262
e CLM temporal-inferior [mm]	0.189

Comments

- Relatively small and tilted optic disc (1) with prominent vessel trunk
- "Step" in the temporal peripapillary retina (2)
- Increased global "rim area" and "rim volume" (3) secondary to significant height differences of temporal and nasal disc borders
- Small "cup/disc area ratio" and normal "linear cup/disc ratio"
- Small cup with shallow excavation (4)
- Normal "cup shape measure" value
- Normal standard reference plane height
- Vital appearance of the peripapillary retinal nerve fiber layer with excellent reflectivity (5)
- Moorfields analysis rates all sectors as "within normal limits"
- Dynamic and slightly asymmetric height profile of the contour line (6) with very small 180° depression (7)
- Both polar peaks ("double humps") reach mean retina height (*)

Original printout with:
a Topography image, OS
b Intensitiy image, OS
c 3D-image, OS

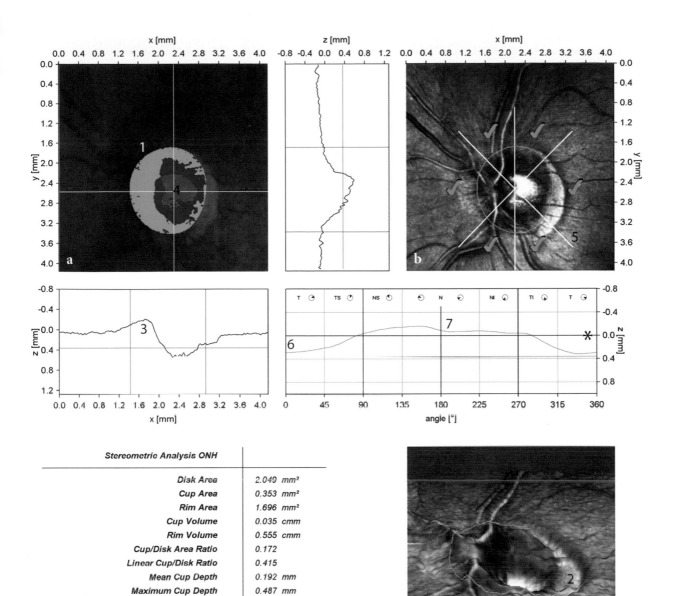

Stereometric Analysis ONH	
Disk Area	2.049 mm²
Cup Area	0.353 mm²
Rim Area	1.696 mm²
Cup Volume	0.035 cmm
Rim Volume	0.555 cmm
Cup/Disk Area Ratio	0.172
Linear Cup/Disk Ratio	0.415
Mean Cup Depth	0.192 mm
Maximum Cup Depth	0.487 mm
Cup Shape Measure	-0.164
Height Variation Contour	0.480 mm
Mean RNFL Thickness	0.313 mm
RNFL Cross Sectional Area	1.588 mm²
Reference Height	0.356 mm
Topography Std Dev.	12 µm

Plate 11

Comments

- Oval, relatively small and tilted (1) optic disc with prominent vessel trunk
- Despite comparison with a fundus photograph, disc borders are difficult to define
- Small cup with shallow excavation
- Vital appearance of the peripapillary retinal nerve fiber layer with excellent reflectivity (2)

a Photograph of the optic disc, OS
b Intensitiy image of interacive measurement, OS

Plate 12

Comments

- Oval tilted optic disc with nasal vessel trunk
- "Step" in the temporal peripapillary retina (1)
- 3D image helps to define disc borders especially in the temporal region
- Large cup with pronounced excavation of oval shape (2)
- Vital appearance of the peripapillary retinal nerve fiber layer with excellent reflectivity (3)

a 3D-image, OD
b Intensitiy image of interacive measurement, OS

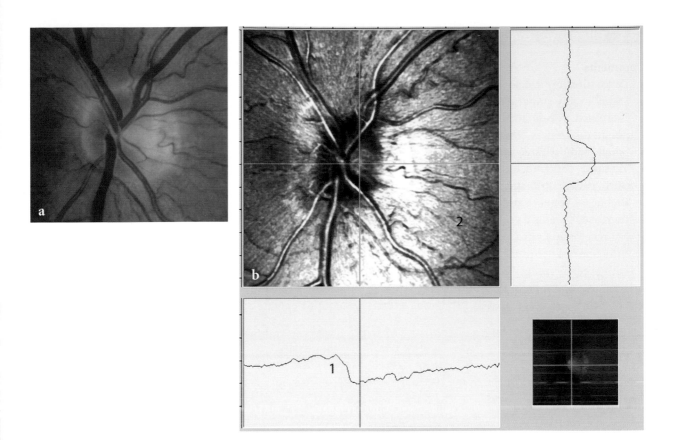

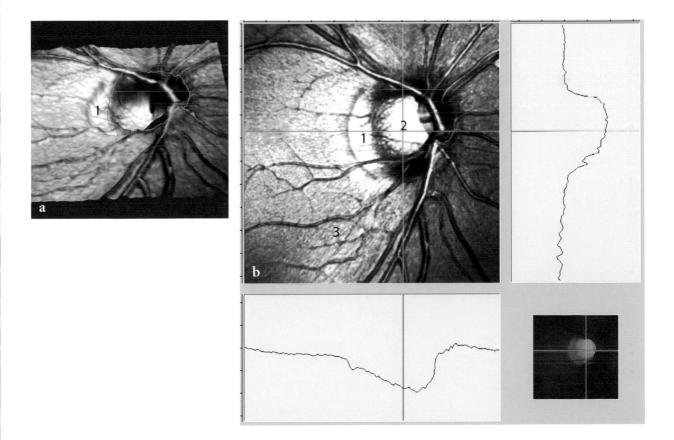

Plate 13

Comments

- Large tilted optic disc (1)
- "Step" in the temporal peripapillary retina (2)
- Increased global "rim area" and "rim volume" seondary to significant height differences of temporal and nasal disc borders (3)
- Small "cup/disc area ratio" and "linear cup/disc ratio"
- Vertically pronounced cup with shallow excavation (4)
- Borderline „cup shape measure" value
- Increased "height variation contour" value
- Increased standard reference plane height (red line)
- Moorfields analysis rates all sectors as "within normal limits"
- Vital appearance of the peripapillary retinal nerve fiber layer with excellent reflectivity (5)
- Dynamic and slightly asymmetric height profile of the contour line (6)
- Superior segment polar peak (7) does not reach mean retina height

Original printout with:
a Topography image, OD
b Intensitiy image, OD
c 3D-image, OD

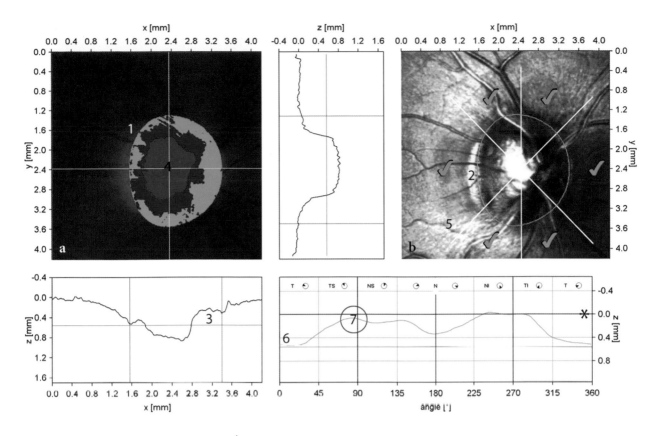

Stereometric Analysis ONH	
Disk Area	3.224 mm²
Cup Area	0.939 mm²
Rim Area	2.286 mm²
Cup Volume	0.159 cmm
Rim Volume	0.764 cmm
Cup/Disk Area Ratio	0.291
Linear Cup/Disk Ratio	0.540
Mean Cup Depth	0.276 mm
Maximum Cup Depth	0.638 mm
Cup Shape Measure	-0.113
Height Variation Contour	0.563 mm
Mean RNFL Thickness	0.333 mm
RNFL Cross Sectional Area	2.122 mm²
Reference Height	0.555 mm
Topography Std Dev.	10 µm

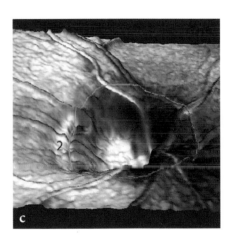

3 Glaucomas

A. F. Scheuerle, E. Schmidt

3.1 Automated Classification Procedures

Laser scanning ophthalmoscopy provides a topographic height map of the optic disc and the peripapillary retina. The obtained data is objective, has a high spatial resolution and reproducibility. Using the Heidelberg Retina Tomograph (HRT) standard software 22 stereometric parameters are presented in a table. Several trials have been reported how HRT parameters can be identified that best separate patients with glaucoma from normal subjects. A major difficulty is that the anatomy of the optic nerve head exhibits a large interindividual variability.

Multivariant analyses were applied to select surface topography parameters that promise a good separation of normal optic discs from glaucomatous ones, irrespective of their clinical appearance. Patients in the glaucoma groups were defined by intraocular pressure >21 mm Hg and reproducible visual field defects. It turned out that a combination of parameters reached higher sensitivity and specificity values than any single parameter. Therefore, several parameters were considered simultaneously to identify glaucomatous structural damage.

In the Mikelberg discriminant function, (FSM) "cup shape measure", "rim volume" and "height variation contour" were combined in a formula taking into account the patient's age. If the resulting value of this calculation is 0, the optic disc will be classified as "normal." A value less than 0 indicates a glaucomatous configuration. The FSM classification is based on HRT parameters which rely on a reference plane. Using the standard reference plane (Burk et al. 2000), a sensitivity of 89% and a specificity of 84% were reported (Mikelberg et al. 1995). Like the FSM, the Burk discriminant function (RB) results in a value which is either positive (normal) or negative (glaucoma); the underlying formula is, however, completely different.

In the course of glaucoma, a remarkable flattening of the optic disc contour line has been observed (Burk et al. 1990). The RB uses the difference between the contour line mean height (CLM) in the temporal quadrant (TQ) of the optic disc, the CLM difference between the temporal superior octant (TSO) and TQ, and the "cup shape measure" (also called "optic disc cup steepness" or "third moment" of frequency distribution of optic cup depth readings) in the TSO. Therefore, it is independent of a reference plane but still dependent from a well drawn contour line (Burk et al. 1998).

As a single parameter, the "retinal nerve fiber layer cross sectional area," the "third moment or cup shape measure" and the global "cup-disc area ratio" could have the greatest diagnostic precision in identifying an abnormal disc (Uchida et al. 1996). However, these parameters are related to disc size, pointed out by studies of Iester and Bathija (Iester et al. 1997, Bathija et al. 1998). There is a strong relationship between the size of the optic disc area and the neuroretinal rim area. It has been reported that in a normal population, the variability of neuroretinal rim area values increases as the neuroretinal rim area itself increases (Britton et al. 1987; Jonas et al. 1988). Therefore, Wollstein et al. defined the normal ranges by using linear regression between the optic disc area and the log of neuroretinal rim area and between the optic disc area and the cup disc area ratio (Wollstein et al. 1998).

Sensitivity was further improved by using the results of separate optic disc segments to detect early, localized defects. In the actual HRT software, the so called "Moorfields regression analysis" calculates 95.0 %, 99.0 % and 99.9 % confidence interval limits for each disc sector and for the global disc, i.e. 95.0 %, 99.0 % or 99.9 % of all optic nerve heads in the normal database have a percentage share of rim area that is greater than this limit. Taking into account optic disc size, this method has the highest specificity and sensitivity values to detect early glaucomatous changes in the optic nerve head reported in the literature. Even in comparison with the clinical assessment of stereoscopic optic nerve head photographs, the HRT image analysis was tested to be more sensitive in distinguishing between healthy persons and patients with early glaucoma (Nakla et al. 1999; Wollstein et al. 2000). Nevertheless, this method is purely a statistical description of whether a disc lies within the defined normal limits. In a recent study, HRT-based classification of "glaucoma" or "normal" was moderately sensitive but not very specific when compared with clinical impression with standard binocular ophthalmoscopy (Kesen et al. 2002). It is also to be noted that all published studies concerning the validity and reliability of discriminant functions or regression analysis are based on relatively small study populations that certainly do not include the natural variety of optic disc samples. The Megalopapilla, for instance, is an entity of relatively high frequency, characterized by a large optic nerve head ($>3.0\,mm^2$) which might appear abnormal due to its increased cup (Sampaolesi et al. 2001). Despite normal rim volume, normal visual field and normal IOP, megalopapillas are often rated as glaucomatous by automated classification procedures including the Moorfields regression analysis.

Furthermore, all these procedures require a contour line with most of them requiring a reference plane as well. The position of the standard reference plane and the majority of stereometric parameters depend on a manual drawn contour line. "Cup shape measure", "maximum cup depth", "height variation contour" and "mean height contour" showed small interobserver variation. "Volume below/above surface" and "volume below/above reference" showed high interobserver variation but could be significantly improved when clinical optic disc slides were additionally available for outlining the optic disc margin (Iester 2001). Paying attention to the correct placement of contour lines, stereometric parameters and automated classification procedures might be highly reproducible with a low operator-induced variability (Garway-Heath et al. 1999; Hatch 1999; Miglior et al. 2002). The three classification procedures mentioned above have been introduced into the standard software of HRT and HRT II as valuable supplements for the clinical evaluation of optic discs and are listed in Table 3.

Table 3 contains two procedures that evaluate surface changes in the peripapillary area. To our knowledge, literature describing the extent of changes in the peripapillary area in the course of glaucoma is rare, in particular concerning a possible relationship with optic disc cupping. Analysis of the geometric profile of the peripapillary area does not require a reference plane and could eliminate variations secondary to imprecise contour line placement. The third method models the shape of the optic nerve head by a smooth two-dimensional surface with a shape described by ten free parameters including the degree of surface curvature of the disc region surrounding the cup, the steepness of the cup walls and measurements of cup width and cup depth without manual outlining of disc boundaries. Up to now the three procedures listed in the table have not been integrated into HRT standard software and are, therefore, reserved for further evaluation on an experimental basis.

Table 3. Statistical evaluation of different automated classification procedures

Reference	Parameters used	Sensitivity (%)	Specificity (%)	Normal eyes (n)	Glaucoma eyes (n)	Visual field: average MD value
Mikelberg et al. 1995[a]	HVC, CSM, RV	87.0	84.4	45	46	−5.5 dB
Iester et al. 1997[a]	HVC, CSM, RV	All eyes: 74.2 DA: <2 mm^2, 64.7; 2–3 mm^2, 78.7; >3 mm^2, 83.3	All eyes: 88.3 DA: <2 mm^2, 83.3; 2–3 mm^2, 89.7; >3 mm^2, 88.9	60	93	−8.3 dB
Uchida et al. 1998[a]	HVC, CSM, RV	80	83		30	−3.7 dB
Burk et al. 1998[b]	CLM (TSO, TQ, TIO), CSM	74.1	85.9	78	58	−15.8 dB
Bathija et al. 1998[c]	HVC, RNFLT, CSM, RA	All eyes: 78.0 DA: <2 mm^2, 71.4; 2–3 mm^2, 94.4	All eyes: 88.0 DA: <2 mm^2, 92.6; 2–3 mm^2, 81.8	49	50	>−10.0 dB
Wollstein et al. 1998[d]	log(RA): CDAR:	84.3 74.5	96.3 97.5	80	51	− 3.6 dB
Caprioli et al. 1998[e]	Peripapillary surface slope	85.0	81.0	43	53	− 4.8 dB
Schmidt, Scheuerle et al. 2003[f]	Peripapillary topography	100.0	100.0	20	20	>−10.0 dB
Swindale et al. 2000[g]	Shape of optic nerve head surface	89.0	89.0	100	100	

DA: disc area
CA: cup area
RA: rim area
CV: cup volume
RV: rim volume
CDAR: cup/disc area ratio
CSM: cup shape measure
HVC: height variation contour
RNFLT: retinal nerve fiber layer thickness
RNFLCA: retinal nerve fiber layer cross sectional area
CLM: contour line mean height
TSO: temporal superior octant
TQ: temporal quadrant
TIO: temporal inferior octant

[a] FSM discriminant function (F≥0: normal; F<0: glaucoma):
F=a–[CSM+0,001981–(50–age (years))]+b–RV+c–HVC+d.
(a=–13,079; b=10,990; c=–7,245; d=–2,662)
(Mikelberg et al. 1995)

[b] RB discriminant function (F≥0: normal; F<0: glaucoma):
F=a–CLM (TIO–TQ)+b–CLM (TSO–TQ)+c–CSM–0,9740553.
(a=4,196594; b=5,641541; c=–3,885066)
(Burk et al. 1998)

[c] Bathija discriminant function (F≥0: normal; F<0: glaucoma):
F=–3.72–4.37–CSM–5.57–HVC+11.78–RNFLT+1.85–RA
(Bathija et al. 1998)

[d] Classification method: Classification result is "glaucoma" if eye falls outside the 99.0% prediction interval of the linear regression of DA with log(RA) and CDAR, respectively, for the global or any predefined segment; otherwise "normal" (Wollstein et al. 1998)

[e] Peripapillary surface slope is radial slope of the retinal surface 0.0–0.25 mm outside the disc margin (Caprioli et al. 1998)

[f] Peripapillary topography is radial slope of the retinal surface height measured in the area 100%–180% of optic disc diameter (Schmidt et al. 2003)

[g] Surface shape is described by ten free parameters acquired without drawing a contour line (Swindale et al. 2000)

Plate 14

Comments

- Very thin neuroretinal rim in the temporal inferior sector (1)
- Decreased global "rim area" and "rim volume"
- Increased "cup/disc area ratio" and "linear cup/disc ratio"
- Vertically pronounced cup shape (2)
- Pathologic "cup shape measure" value
- Moorfields analysis rates only the temporal inferior sector as "out of normal limits"
- Reflectivity image reveals a nerve fiber bundle defect in the temporal inferior sector (3)
- Flattened height profile of the contour line (4), with depression in the temporal inferior sector (5)
- Both polar peaks ("double humps") (6) do not reach mean retina height (*)

Case: A 46 year old female with primary open angle glaucoma and visual field defects (MD 4.4 dB).

Original printout with:
a Topography image, OS
b Intensitiy image, OS
c Photograph of the optic disc, OS

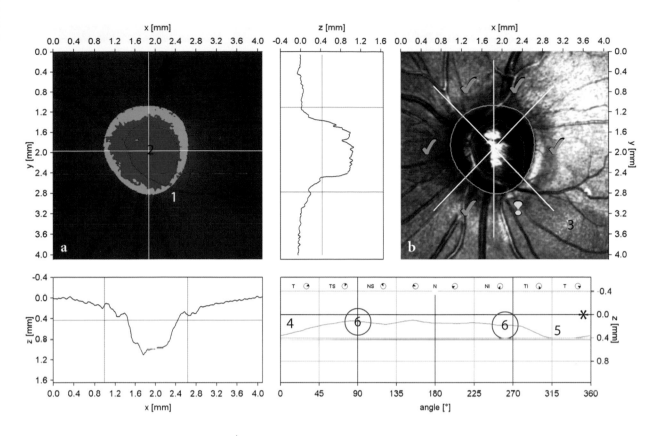

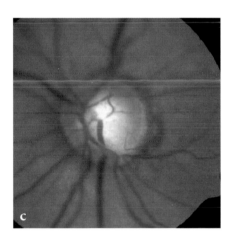

Stereometric Analysis ONH	
Disk Area	2.203 mm²
Cup Area	0.913 mm²
Rim Area	1.290 mm²
Cup Volume	0.342 cmm
Rim Volume	0.278 cmm
Cup/Disk Area Ratio	0.414
Linear Cup/Disk Ratio	0.644
Mean Cup Depth	0.375 mm
Maximum Cup Depth	0.820 mm
Cup Shape Measure	-0.077
Height Variation Contour	0.322 mm
Mean RNFL Thickness	0.217 mm
RNFL Cross Sectional Area	1.141 mm²
Reference Height	0.429 mm
Topography Std Dev.	16 µm

Plate 15

Comments

- Very thin neuroretinal rim in the temporal superior sector (1)
- Pathologically decreased global "rim area" and "rim volume"
- Pathologically increased "cup/disc area ratio" and "linear cup/disc ratio"
- Circular cup with deep excavation
- Pathologically decreased "height variation contour" value
- Pathologic "cup shape measure" value
- Moorfields analysis rates all sectors exept the temporal inferior one as "out of normal limits"
- Both discriminat functions rate optic disc as "glaucomatous"
- Decreased reflectiviy of the peripapillary retinal nerve fiber layer (2)

a 3D-image, OD
b Topography image, OD
c Intensitiy image, OD
d Stereometric parameters
e Photograph of the optic disc, OD

Case: 44 year old female with primary open angle glaucoma presenting with high intraocular pressure (approximately 24 mmHg) and visual field defects (MD 5.2 dB) despite triple antiglaucomatous therapy.

Plate 16

Comments

- A thinned neuroretinal rim in the temporal superior sector (1)
- Pathologically decreased global "rim area" and "rim volume"
- Pathologically increased "cup/disc area ratio" and "linear cup/disc ratio"
- Circular cup with deep excavation (2)
- Pathologically decreased "height variation contour" value
- Pathologic "cup shape measure" value
- Moorfields analysis rates all sectors except the temporal one as "out of normal limits"
- Both discriminat functions rate optic disc as "glaucomatous"
- Decreased reflectiviy of the peripapillary retinal nerve fiber layer (3)

a 3D-image, OS
b Topography image, OS
c Intensitiy image, OS
d Stereometric parameters
e Photograph of the optic disc, OS

Note the symmetric development of glaucomatous damage in the right (upper plate) and left optic disc of this patient.

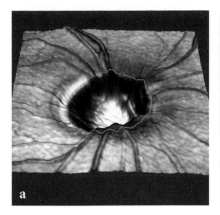

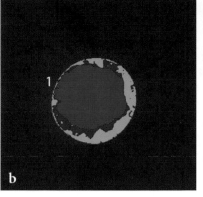

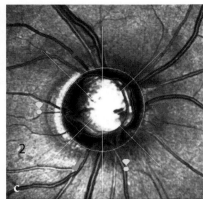

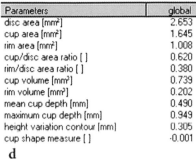

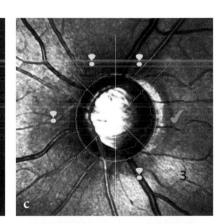

Parameters	global		Parameters	global
disc area [mm²]	2.653		mean RNFL thickness [mm]	0.175
cup area [mm²]	1.645		RNFL cross sectional area [mm²]	1.012
rim area [mm²]	1.008		linear cup/disc ratio [ ]	0.787
cup/disc area ratio [ ]	0.620		maximum contour elevation [mm]	0.038
rim/disc area ratio [ ]	0.380		maximum contour depression [mm]	0.343
cup volume [mm²]	0.739		CLM temporal-superior [mm]	0.078
rim volume [mm²]	0.202		CLM temporal-inferior [mm]	0.161
mean cup depth [mm]	0.490		average variability (SD) [mm]	0.011
maximum cup depth [mm]	0.949		reference height [mm]	0.376
height variation contour [mm]	0.305		FSM discriminant function value [ ]	-2.643
cup shape measure [ ]	-0.001		RB discriminant function value [ ]	-0.251

d

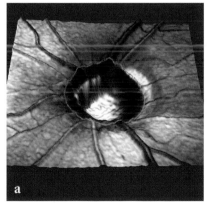

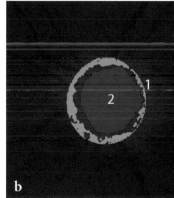

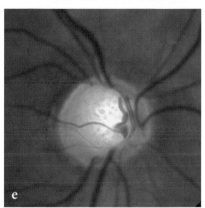

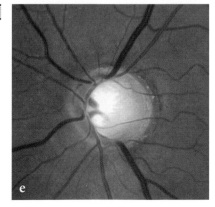

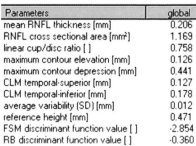

Parameters	global		Parameters	global
disc area [mm²]	2.565		mean RNFL thickness [mm]	0.206
cup area [mm²]	1.474		RNFL cross sectional area [mm²]	1.169
rim area [mm²]	1.091		linear cup/disc ratio [ ]	0.758
cup/disc area ratio [ ]	0.575		maximum contour elevation [mm]	0.126
rim/disc area ratio [ ]	0.425		maximum contour depression [mm]	0.441
cup volume [mm²]	0.707		CLM temporal-superior [mm]	0.127
rim volume [mm²]	0.224		CLM temporal-inferior [mm]	0.178
mean cup depth [mm]	0.507		average variability (SD) [mm]	0.012
maximum cup depth [mm]	0.967		reference height [mm]	0.471
height variation contour [mm]	0.315		FSM discriminant function value [ ]	-2.854
cup shape measure [ ]	0.028		RB discriminant function value [ ]	-0.360

d

Comments

- Very thin neuroretinal rim in the temporal/temporal inferior sector (1)
- Pathologically decreased global "rim area" and "rim volume"
- Pathologically increased "cup/disc area ratio" and "linear cup/disc ratio"
- Vertically pronounced cup shape with deep excavation (2)
- Normal "height variation contour" value
- Borderline "cup shape measure" value
- Moorfields analysis rates only the temporal and nasal inferior sector as "within normal limits"
- Flattened height profile of the contour line (3) with some depression in the temporal inferior sector (4)
- Both polar peaks ("double humps") (5) do not reach mean retina height (*)

Original printout with:
a Topography image, OD
b Intensitiy image, OD
c 30°-visual field, OD

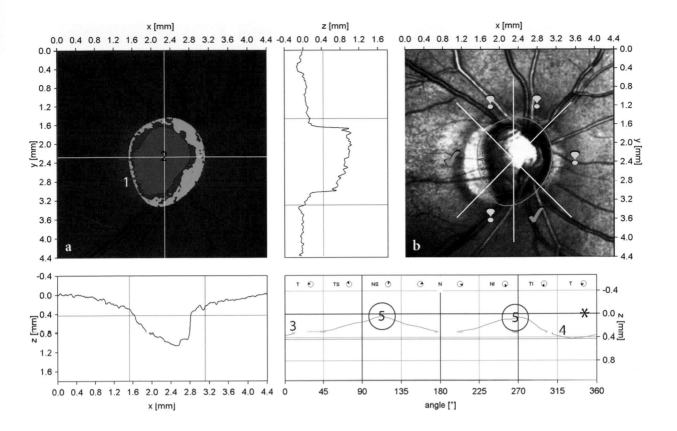

Stereometric Analysis ONH	
Disk Area	2.303 mm²
Cup Area	1.220 mm²
Rim Area	1.083 mm²
Cup Volume	0.441 cmm
Rim Volume	0.252 cmm
Cup/Disk Area Ratio	0.530
Linear Cup/Disk Ratio	0.728
Mean Cup Depth	0.421 mm
Maximum Cup Depth	1.033 mm
Cup Shape Measure	-0.107
Height Variation Contour	0.380 mm
Mean RNFL Thickness	0.202 mm
RNFL Cross Sectional Area	1.088 mm²
Reference Height	0.431 mm
Topography Std Dev.	20 µm

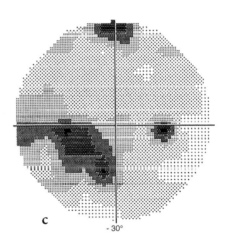

Plate 18

Comments

- Very thin neuroretinal rim in the temporal sector (1)
- Pathologically decreased global "rim area" and "rim volume"
- Pathologically increased "cup/disc area ratio" and "linear cup/disc ratio"
- Vertically pronounced cup shape
- Decreased "height variation contour" value
- Borderline "cup shape measure" value
- Moorfields analysis rates only the nasal and nasal superior sector as "within normal limits"
- Both discriminat functions rate optic disc as "glaucomatous"

a 30°-visual field, OD
b Topography image, OD
c Intensitiy image, OD
d Stereometric parameters

Plate 19

Comments

- Very thin neuroretinal rim in the temporal/temporal superior sector (1)
- Pathologically decreased global "rim area" and "rim volume"
- Pathologically increased "cup/disc area ratio" and "linear cup/disc ratio"
- Vertically pronounced cup shape
- Pathologically decreased "height variation contour" value
- Pathologic "cup shape measure" value
- Moorfields analysis rates only two superior sectors as "out of normal limits"
- Both discriminat functions rate optic disc as "glaucomatous"

Note the similar development of glaucomatous damage in the right (upper plate) and left optic disc of this patient.

a 30°-visual field, OS
b Topography image, OS
c Intensitiy image, OS
d Stereometric parameters

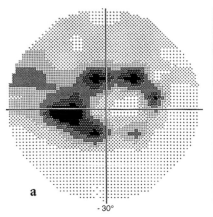

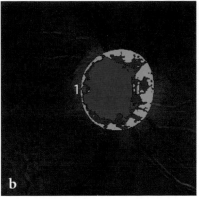

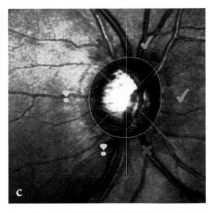

Parameters	global		Parameters	global
disc area [mm²]	2.128		mean RNFL thickness [mm]	0.187
cup area [mm²]	1.117		RNFL cross sectional area [mm²]	0.966
rim area [mm²]	1.011		linear cup/disc ratio [ ]	0.724
cup/disc area ratio [ ]	0.525		maximum contour elevation [mm]	0.012
rim/disc area ratio [ ]	0.475		maximum contour depression [mm]	0.367
cup volume [mm²]	0.292		CLM temporal-superior [mm]	0.111
rim volume [mm²]	0.192		CLM temporal-inferior [mm]	0.088
mean cup depth [mm]	0.321		average variability (SD) [mm]	0.039
maximum cup depth [mm]	0.757		reference height [mm]	0.402
height variation contour [mm]	0.355		FSM discriminant function value [ ]	-1.191
cup shape measure [ ]	-0.111		RB discriminant function value [ ]	-0.161

(d)

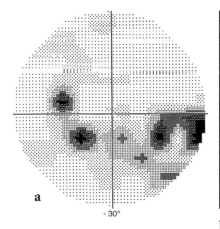

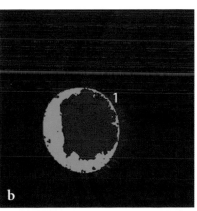

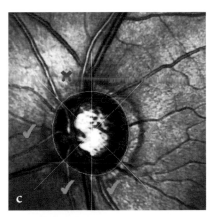

Parameters	global		Parameters	global
disc area [mm²]	2.211		mean RNFL thickness [mm]	0.113
cup area [mm²]	1.152		RNFL cross sectional area [mm²]	0.594
rim area [mm²]	1.059		linear cup/disc ratio [ ]	0.722
cup/disc area ratio [ ]	0.521		maximum contour elevation [mm]	0.076
rim/disc area ratio [ ]	0.479		maximum contour depression [mm]	0.300
cup volume [mm²]	0.313		CLM temporal-superior [mm]	0.042
rim volume [mm²]	0.190		CLM temporal-inferior [mm]	0.046
mean cup depth [mm]	0.313		average variability (SD) [mm]	0.036
maximum cup depth [mm]	0.690		reference height [mm]	0.305
height variation contour [mm]	0.224		FSM discriminant function value [ ]	-1.003
cup shape measure [ ]	-0.054		RB discriminant function value [ ]	-0.727

(d)

Comments

- Medium-sized tilted disc (1)
- Very thin neuroretinal rim in the temporal/temporal inferior sector (2)
- Decreased global "rim area" and "rim volume"
- Pathologically increased "cup/disc area ratio" and "linear cup/disc ratio"
- Vertically pronounced cup shape with shallow excavation (3)
- Decreased "height variation contour" value
- Pathologic "cup shape measure" value
- Moorfields analysis rates only the temporal inferior sector as "out of normal limits"
- Decreased reflectivity of the peripapillary retinal nerve fiber layer (4)
- Slightly flattened height profile of the contour line (5)
- Both polar peaks ("double humps") (6) just reach mean retina height (*)

Case: A 62 year old female with recently diagnosed primary open angle glaucoma and beginning visual field defects (MD 3.1 dB).

Original printout with:

a Topography image, OS
b Intensitiy image, OS
c Photograph of the optic disc, OS

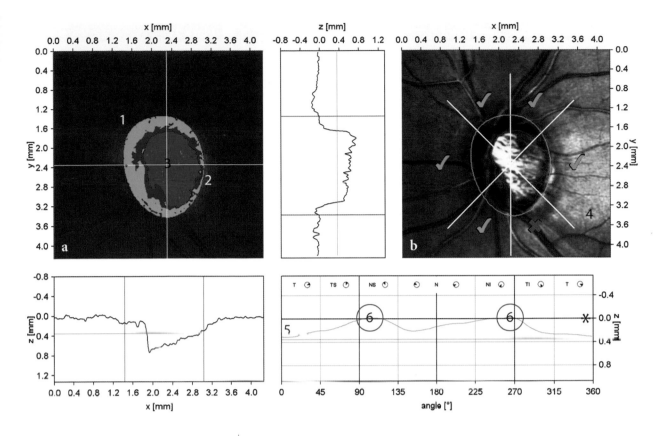

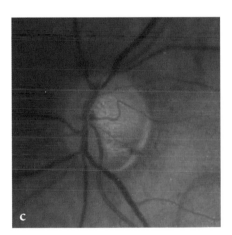

Stereometric Analysis ONH	
Disk Area	2.601 mm²
Cup Area	1.222 mm²
Rim Area	1.379 mm²
Cup Volume	0.222 cmm
Rim Volume	0.349 cmm
Cup/Disk Area Ratio	0.470
Linear Cup/Disk Ratio	0.686
Mean Cup Depth	0.289 mm
Maximum Cup Depth	0.648 mm
Cup Shape Measure	-0.077
Height Variation Contour	0.333 mm
Mean RNFL Thickness	0.216 mm
RNFL Cross Sectional Area	1.237 mm²
Reference Height	0.352 mm
Topography Std Dev.	15 µm

Plate 21

Comments

- Very thin neuroretinal rim in the temporal/temporal inferior sector (1)
- Edema map confirms the thinned neuroretinal rim (2)
- Pathologically decreased global "rim area" and "rim volume"
- Pathologically increased "cup/disc area ratio" and "linear cup/disc ratio"
- Vertically pronounced cup shape with deep excavation (3)
- Pathologically decreased "height variation contour" value
- Pathologic "cup shape measure" value
- Moorfields analysis rates the two inferior sectors and the temporal superior sector as "out of normal limits"
- Both discriminat functions rate optic disc as "glaucomatous"
- Flattened height profile of the contour line (4) with depression in the temporal inferior sector (5)
- Both polar peaks ("double humps") (6) do not reach mean retina height (✳)

a   Topography image, OD
b   Edema map, OD
c   Intensity image, OD
d   Profile of contour line
e   Stereometric parameters

Plate 22

Comments

- Very thin neuroretinal rim in the temporal/temporal superior sector (1)
- Pathologically decreased global "rim area" and "rim volume"
- Pathologically increased "cup/disc area ratio" and "linear cup/disc ratio"
- Circular cup shape with deep excavation (2)
- Pathologically decreased "height variation contour" value
- Pathologic "cup shape measure" value
- Moorfields analysis rates all sectors except the temporal inferior one as "out of normal limits"
- Both discriminat functions rate optic disc as "glaucomatous"
- Reflectivity images (b, c) demonstrate two nerve fiber bundle defects in the temporal sectors (3), more visible in the black and white mode (4)

a   Topography image, OD
b   Intensity image, b/w, OD
c   Intensity image, OD
d   Stereometric parameters

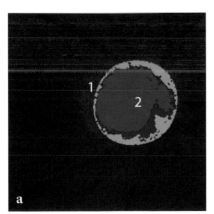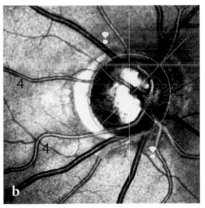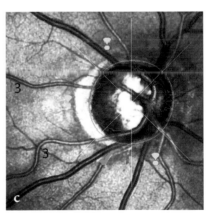

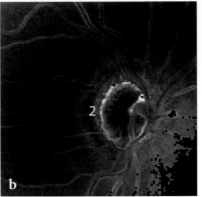

d

Parameters	global
disc area [mm²]	2.242
cup area [mm²]	1.181
rim area [mm²]	1.060
cup/disc area ratio [ ]	0.527
rim/disc area ratio [ ]	0.473
cup volume [mm³]	0.476
rim volume [mm³]	0.220
mean cup depth [mm]	0.439
maximum cup depth [mm]	0.798
height variation contour [mm]	0.255
cup shape measure [ ]	0.023
mean RNFL thickness [mm]	0.193
RNFL cross sectional area [mm²]	1.024
linear cup/disc ratio [ ]	0.726
maximum contour elevation [mm]	0.092
maximum contour depression [mm]	0.347
CLM temporal-superior [mm]	0.128
CLM temporal-inferior [mm]	0.086
average variability (SD) [mm]	0.023
reference height [mm]	0.387
FSM discriminant function value [ ]	-1.985
e RB discriminant function value [ ]	-0.784

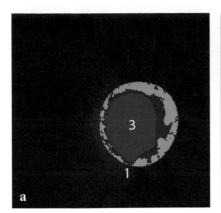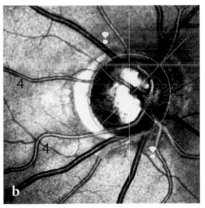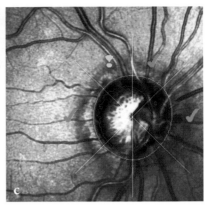

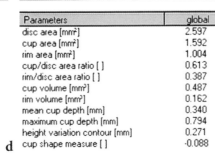

Parameters	global	Parameters	global
disc area [mm²]	2.597	mean RNFL thickness [mm]	0.133
cup area [mm²]	1.592	RNFL cross sectional area [mm²]	0.761
rim area [mm²]	1.004	linear cup/disc ratio [ ]	0.783
cup/disc area ratio [ ]	0.613	maximum contour elevation [mm]	0.048
rim/disc area ratio [ ]	0.387	maximum contour depression [mm]	0.319
cup volume [mm³]	0.487	CLM temporal-superior [mm]	0.092
rim volume [mm³]	0.162	CLM temporal-inferior [mm]	0.085
mean cup depth [mm]	0.340	average variability (SD) [mm]	0.033
maximum cup depth [mm]	0.794	reference height [mm]	0.356
height variation contour [mm]	0.271	FSM discriminant function value [ ]	-1.907
d cup shape measure [ ]	-0.088	RB discriminant function value [ ]	-0.260

Comments

- Very thin neuroretinal rim in the temporal/temporal inferior sector (1)
- Pathologically decreased global "rim area" and "rim volume"
- Pathologically increased "cup/disc area ratio" and "linear cup/disc ratio"
- Vertically pronounced cup shape (2)
- Pathologically decreased "height variation contour" value
- Borderline "cup shape measure" value
- Moorfields analysis rates only the temporal inferior sector as "out of normal limits"
- The reflectivity image demonstrates a nerve fiber bundle defect in the temporal inferior sector (3)
- Corresponding visual field defect (4)
- Severly flattened height profile of the contour line (5)
- Both polar peaks ("double humps") (6) are far below mean retina height (∗)
- Corresponding visual field defect (MD 5.3 dB) (d)

Original printout with:
a Topography image, OD
b Intensity image, OD
c Moorfields classification
d 30°-visual field, OD

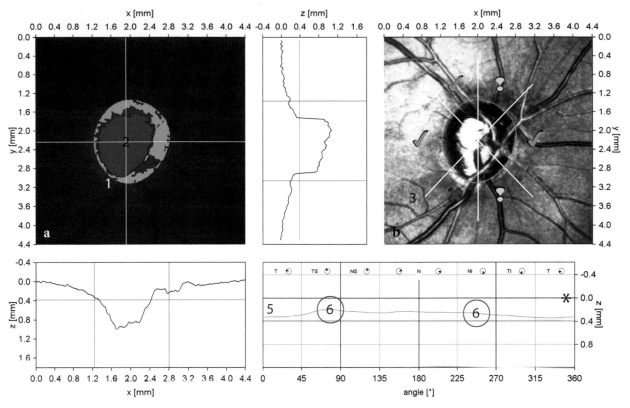

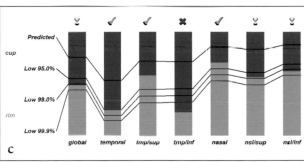

Stereometric Analysis ONH

Disk Area	2.075 mm²
Cup Area	1.060 mm²
Rim Area	1.015 mm²
Cup Volume	0.399 cmm
Rim Volume	0.140 cmm
Cup/Disk Area Ratio	0.511
Linear Cup/Disk Ratio	0.715
Mean Cup Depth	0.354 mm
Maximum Cup Depth	0.776 mm
Cup Shape Measure	-0.104
Height Variation Contour	0.148 mm
Mean RNFL Thickness	0.109 mm
RNFL Cross Sectional Area	0.560 mm²
Reference Height	0.382 mm
Topography Std Dev.	11 μm

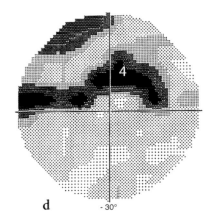

Comments

- Oval disc shape
- Pathologically decreased global "rim volume" but normal "rim area"
- Normal "cup/disc area ratio" and "linear cup/disc ratio"
- Vertically pronounced cup shape with relatively shallow excavation
- Pathologically decreased "height variation contour" value
- Borderline "cup shape measure" value
- FSM discriminat function rates optic disc as "glaucomatous", RB as "normal"
- Slightly flattened height profile of the contour line (1)
- Both polar peaks ("double humps") (2) do not reach mean retina height (*)
- The 20°-reflectivity images (b, d) demonstrate a large nerve fiber bundle defect (3) in the temporal inferior sector, more visible in the black and white mode (d)

Plate 24

a Profile of contour line, OD
b 20°-intensity image, OD
c Stereometric parameters
d 20°-intensity image, b/w, OD

Comments

- Thinned neuroretinal rim in the temporal inferior sector (1)
- Vertically pronounced cup shape (2)
- Moorfields analysis rates all sectors except the temporal one as "out of normal limits"
- The reflectivity images (b, c) demonstrate nerve fiber bundle defect (3) in the temporal inferior sector, more visible in the black and white mode (b)
- Corresponding visual field defect (MD 4.7 dB) (4)

Plate 25

a Topography image, OS
b Intensity image, b/w, OS
c Intensity image, OS
d 30°-visual field, OS
e Photograph of the optic disc, OS

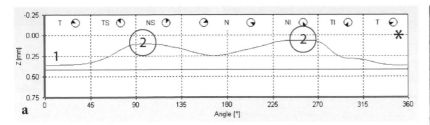

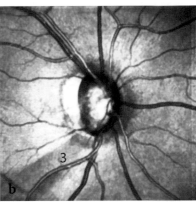

Parameters	global
disc area [mm²]	2.244
cup area [mm²]	0.681
rim area [mm²]	1.562
cup/disc area ratio [ ]	0.304
rim/disc area ratio [ ]	0.696
cup volume [mm³]	0.051
rim volume [mm³]	0.263
mean cup depth [mm]	0.161
maximum cup depth [mm]	0.429
height variation contour [mm]	0.303
cup shape measure [ ]	-0.130
mean RNFL thickness [mm]	0.204
RNFL cross sectional area [mm²]	1.089
linear cup/disc ratio [ ]	0.551
maximum contour elevation [mm]	0.063
maximum contour depression [mm]	0.367
CLM temporal-superior [mm]	0.138
CLM temporal-inferior [mm]	0.130
average variability (SD) [mm]	0.009
reference height [mm]	0.418
FSM discriminant function value [ ]	-0.577
RB discriminant function value [ ]	0.493

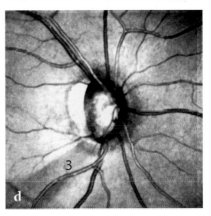

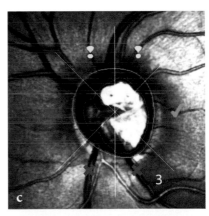

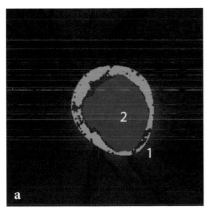

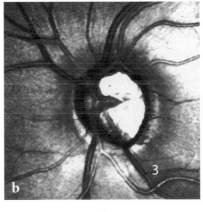

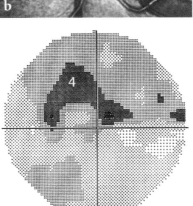

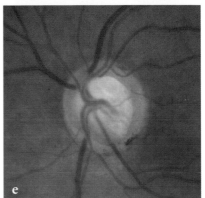

Plate 26

Comments

- Thinned neuroretinal rim in the temporal sector (1)
- Vertically pronounced cup shape (2)
- Vessel trunk simulates normal neuroretinal rim in the nasal hemisphere of the optic disc (3)
- Vessel kinking in the temporal inferior sector (4)
- The reflectivity images (b, c) demonstrate a nerve fiber bundle defect (5) in the temporal inferior sector, more visible in the black and white mode (b)
- Corresponding visual field defect (MD 2.7 dB) (d)

a Topography image, OS
b Intensityimage, b/w, OS
c Intensity image, OS
d 30°-visual field, OS
e Photograph of the optic dics, OS

Plate 27

Comments

- Very thin neuroretinal rim in the temporal inferior sector (1)
- Pathologically decreased global "rim volume" and "rim area"
- Pathologically increased "cup/disc area ratio" and "linear cup/disc ratio"
- Vertically pronounced cup shape (2)
- Pathologically decreased "height variation contour" value
- Pathologic "cup shape measure" value
- Both discriminat functions rate optic disc as "glaucomatous"
- The reflectivity images (b, c) demonstrate a nerve fiber bundle defect in the temporal inferior sector (3), more visible in the black and white mode (b)
- Corresponding visual field defect (MD 5.0 dB) (d)

a Topography image, OS
b Intensity image, b/w, OS
c Intensity image, OS
d 30°-visual field, OS
e Stereometric parameters

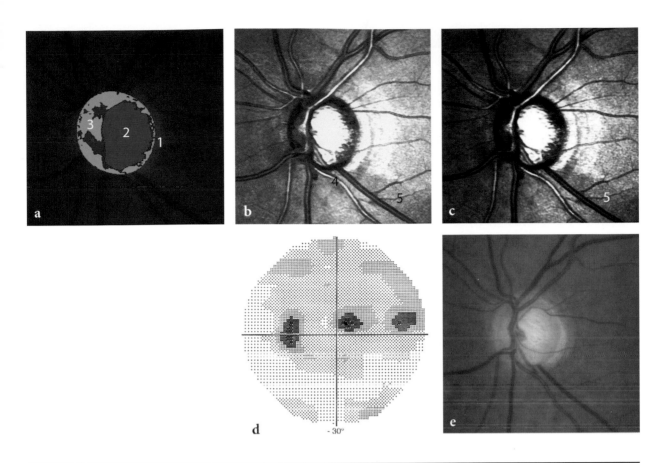

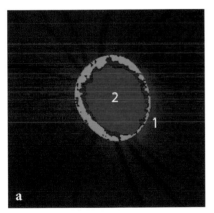

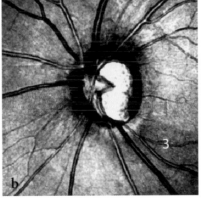

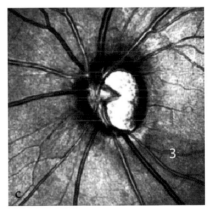

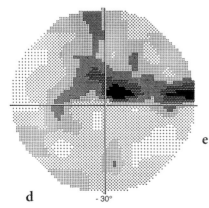

Parameters	global
disc area [mm²]	2.030
cup area [mm²]	1.276
rim area [mm²]	0.754
cup/disc area ratio [ ]	0.628
rim/disc area ratio [ ]	0.372
cup volume [mm³]	0.436
rim volume [mm³]	0.087
mean cup depth [mm]	0.359
maximum cup depth [mm]	0.712
height variation contour [mm]	0.097
cup shape measure [ ]	-0.020

Parameters	global
mean RNFL thickness [mm]	0.103
RNFL cross sectional area [mm²]	0.523
linear cup/disc ratio [ ]	0.793
maximum contour elevation [mm]	0.167
maximum contour depression [mm]	0.264
CLM temporal-superior [mm]	0.069
CLM temporal-inferior [mm]	0.040
average variability (SD) [mm]	0.008
reference height [mm]	0.308
FSM discriminant function value [ ]	-1.721
RB discriminant function value [ ]	-1.260

Comments

- Pathologically decreased global "rim area" and "rim volume"
- Pathologically increased "cup/disc area ratio" and "linear cup/disc ratio"
- Horizontally pronounced cup shape (1)
- Pathologically decreased "height variation contour" value
- Pathologic "cup shape measure" value
- Moorfields analysis rates only the temporal and temporal inferior sector as "within normal limits"
- Decreased retinal reflectivity (2)
- Severly flattened height profile of the contour line (3)
- Both polar peaks ("double humps") (4) are far below mean retina height (∗)

Original printout with:
a Topography image, OD
b Intensity image, OD
c Photograph of the optic disc, OD

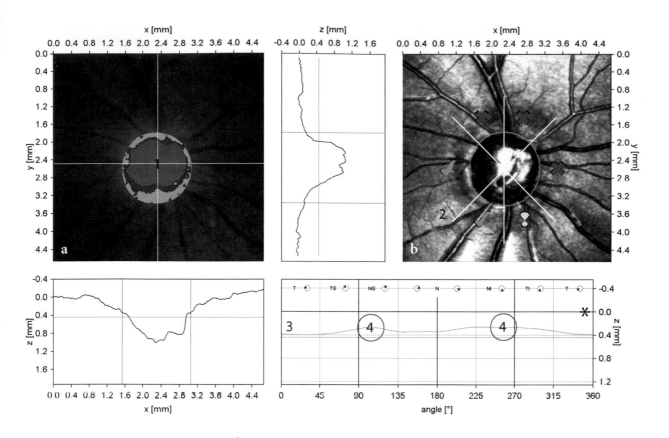

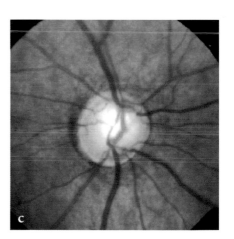

Stereometric Analysis ONH	
Disk Area	1.953 mm²
Cup Area	1.090 mm²
Rim Area	0.863 mm²
Cup Volume	0.333 cmm
Rim Volume	0.111 cmm
Cup/Disk Area Ratio	0.558
Linear Cup/Disk Ratio	0.747
Mean Cup Depth	0.313 mm
Maximum Cup Depth	0.701 mm
Cup Shape Measure	-0.099
Height Variation Contour	0.141 mm
Mean RNFL Thickness	0.113 mm
RNFL Cross Sectional Area	0.558 mm²
Reference Height	0.440 mm
Topography Std Dev.	9 µm

Comments

- Medium-sized optic disc (1)
- Nearly complete loss of the neuroretinal rim (2)
- Pathologically decreased global "rim area" and "rim volume"
- Pathologically increased "cup/disc area ratio" and "linear cup/disc ratio"
- Very large circular cup with very deep excavation (3)
- Pathologically decreased "height variation contour" value
- Pathologic "cup shape measure" value
- Moorfields analysis rates all sectors as "out of normal limits"
- Severly decreased retinal reflectivity (4)
- Severly flattened height profile of the contour line (5)
- Both polar peaks ("double humps") (6) are far below mean retina height (*)

Original printout with:
a Topography image, OS
b Intensity image, OS
c Photograph of the optic disc, OS

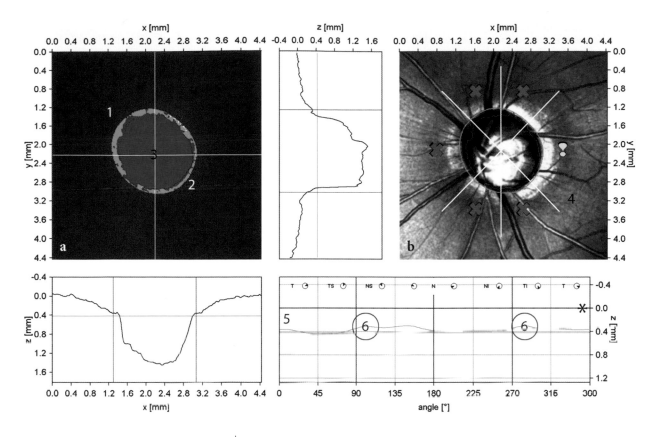

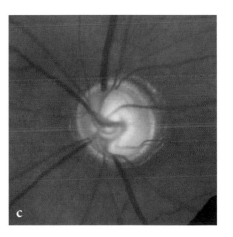

Stereometric Analysis ONH	
Disk Area	2.467 mm²
Cup Area	1.938 mm²
Rim Area	0.529 mm²
Cup Volume	1.277 cmm
Rim Volume	0.040 cmm
Cup/Disk Area Ratio	0.786
Linear Cup/Disk Ratio	0.886
Mean Cup Depth	0.655 mm
Maximum Cup Depth	1.079 mm
Cup Shape Measure	0.130
Height Variation Contour	0.138 mm
Mean RNFL Thickness	0.053 mm
RNFL Cross Sectional Area	0.295 mm²
Reference Height	0.417 mm
Topography Std Dev.	9 μm

Plate 30

Comments

- Medium-sized optic disc (1)
- Nearly complete loss of the neuroretinal rim (2)
- Pathologically decreased global "rim area" and "rim volume"
- Pathologically increased "cup/disc area ratio"
  and "linear cup/disc ratio"
- Very large circular cup (3) with very deep excavation (4) also shown
  in the 3D-reconstruction (d)
- Pathologically decreased "height variation contour" value
- Pathologic "cup shape measure" value
- Moorfields analysis rates all sectors as "out of normal limits"
- Severly decreased retinal reflectivity (5)
- Severly flattened height profile of the contour line (6)
- Both polar peaks ("double humps") (7) are far below mean retina
  height (*)

Original printout with:
a Topography image, OD
b Intensity image, OD
c Moorfields classification
d 3D-image, OD

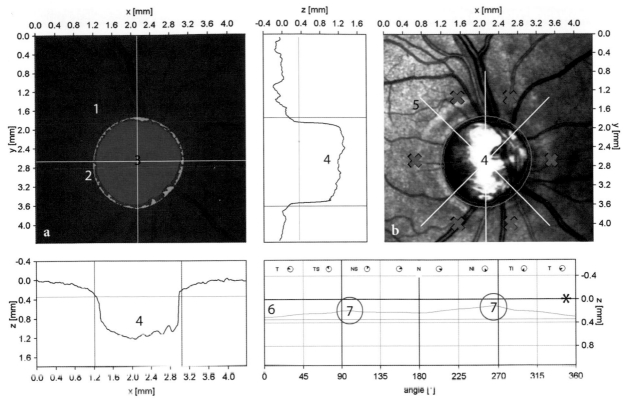

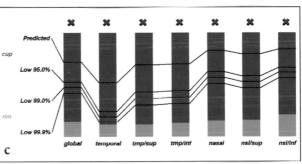

Stereometric Analysis ONH	
Disk Area	2.683 mm²
Cup Area	2.285 mm²
Rim Area	0.398 mm²
Cup Volume	1.612 cmm
Rim Volume	0.045 cmm
Cup/Disk Area Ratio	0.852
Linear Cup/Disk Ratio	0.923
Mean Cup Depth	0.747 mm
Maximum Cup Depth	1.089 mm
Cup Shape Measure	0.172
Height Variation Contour	0.107 mm
Mean RNFL Thickness	0.121 mm
RNFL Cross Sectional Area	0.705 mm²
Reference Height	0.340 mm
Topography Std Dev.	24 µm

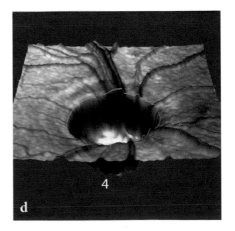

Plate 31

Comments

- Nearly complete loss of the neuroretinal rim in the temporal sectors (1)
- Vessel trunk simulates normal neuroretinal rim in the nasal hemisphere of the optic disc (2)
- Circular peripillary atrophy zone (3)
- Pathologically decreased global "rim area" and "rim volume"
- Pathologically increased "cup/disc area ratio" and "linear cup/disc ratio"
- Very large vertically pronounced cup (4)
- Pathologically decreased "height variation contour" value
- Pathologic "cup shape measure" value
- Moorfields analysis rates all sectors, except the nasal and nasal inferior one, as "out of normal limits"
- Severely decreased retinal reflectivity (5)
- Severely flattened height profile of the contour line (6)
- Both polar peaks ("double humps") (7) are far below mean retina height (*)

Original printout with:
a Topography image, OS
b Intensity image, OS
c 3D-image, OS

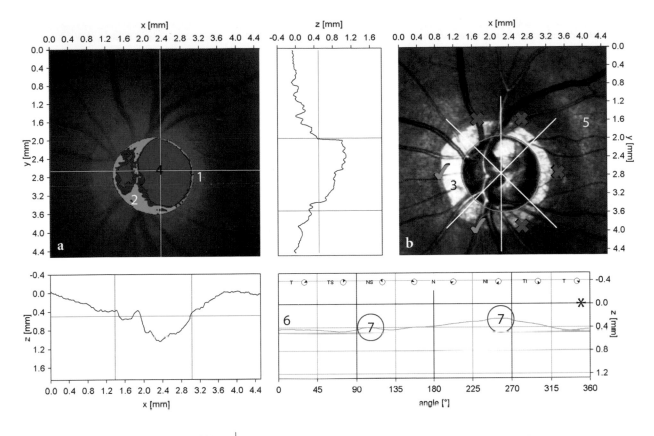

Stereometric Analysis ONH	
Disk Area	2.083 mm²
Cup Area	1.291 mm²
Rim Area	0.793 mm²
Cup Volume	0.383 cmm
Rim Volume	0.104 cmm
Cup/Disk Area Ratio	0.620
Linear Cup/Disk Ratio	0.787
Mean Cup Depth	0.312 mm
Maximum Cup Depth	0.630 mm
Cup Shape Measure	-0.017
Height Variation Contour	0.222 mm
Mean RNFL Thickness	0.101 mm
RNFL Cross Sectional Area	0.516 mm²
Reference Height	0.491 mm
Topography Std Dev.	18 μm

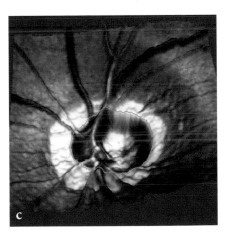

Comments

- Small optic disc (1)
- Nearly complete loss of the neuroretinal rim in the temporal sectors (2)
- Vessel trunk simulates normal neuroretinal rim in the nasal hemisphere of the optic disc (3)
- Pathologically decreased global "rim area" and "rim volume"
- Pathologically increased "cup/disc area ratio" and "linear cup/disc ratio"
- Vertically pronounced cup shape (4)
- Pathologically decreased "height variation contour" value
- Pathologic "cup shape measure" value
- Moorfields analysis rates all sectors, except the nasal inferior one, as "out of normal limits"
- Decreased retinal reflectivity (5)
- Flattened height profile of the contour line (6)
- Both polar peaks ("double humps") (7) do not reach mean retina height (*)

Original printout with:

a Topography image, OD
b Intensity image, OD
c Moorfields classification

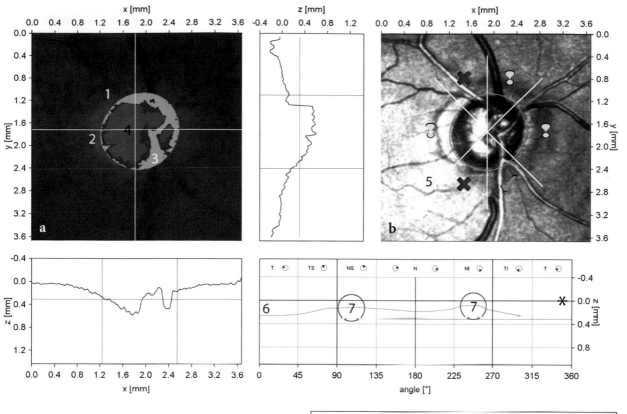

Stereometric Analysis ONH	
Disk Area	1.345 mm²
Cup Area	0.669 mm²
Rim Area	0.676 mm²
Cup Volume	0.086 cmm
Rim Volume	0.112 cmm
Cup/Disk Area Ratio	0.497
Linear Cup/Disk Ratio	0.705
Mean Cup Depth	0.188 mm
Maximum Cup Depth	0.409 mm
Cup Shape Measure	-0.036
Height Variation Contour	0.197 mm
Mean RNFL Thickness	0.127 mm
RNFL Cross Sectional Area	0.523 mm²
Reference Height	0.311 mm
Topography Std Dev.	14 µm

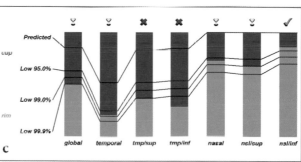

Plate 33

Comments

- Small optic disc (1)
- Very thin neuroretinal rim in the temporal/temporal superior sectors (2)
- Vessel trunk simulates normal neuroretinal rim in the nasal/nasal inferior sector
- Pathologically decreased global "rim area" and "rim volume"
- Pathologically increased "cup/disc area ratio" and "linear cup/disc ratio"
- Vertically pronounced cup shape (3)
- Decreased "height variation contour" value
- Borderline "cup shape measure" value
- Moorfields analysis rates only the two superior sectors as "out of normal limits"
- Temporal peripapillary atrophy (4)
- Decreased retinal reflectivity (5)
- Asymmetric height profile of the contour line (6), flattened in the superior sectors (7)
- The superior segment polar peak (7) does not reach mean retina height (*)
- Corresponding visual field defect (MD 8.3 dB) (c)

Original printout with:
a Topography image, OS
b Intensitiy image, OS
c 30°-visual field, OS

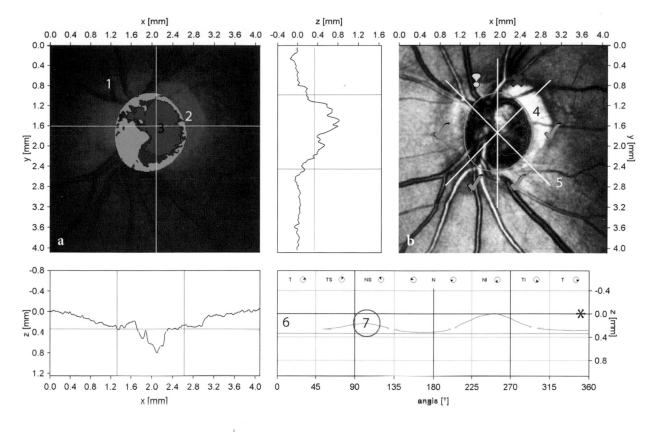

Stereometric Analysis ONH	
Disk Area	1.608 mm²
Cup Area	0.632 mm²
Rim Area	0.976 mm²
Cup Volume	0.105 cmm
Rim Volume	0.162 cmm
Cup/Disk Area Ratio	0.393
Linear Cup/Disk Ratio	0.627
Mean Cup Depth	0.196 mm
Maximum Cup Depth	0.540 mm
Cup Shape Measure	-0.149
Height Variation Contour	0.324 mm
Mean RNFL Thickness	0.118 mm
RNFL Cross Sectional Area	0.533 mm²
Reference Height	0.333 mm
Topography Std Dev.	17 µm

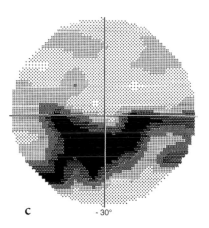

c   - 30°

Plate 34

Comments

- Small optic disc (1)
- Very thin neuroretinal rim in the temporal/temporal inferior sector (2)
- Pathologically decreased global "rim volume" and "rim area"
- Pathologically increased "cup/disc area ratio" and "linear cup/disc ratio"
- Pathologically decreased "height variation contour" value
- Pathologic "cup shape measure" value
- Both discriminat functions rate optic disc as "glaucomatous"
- Moorfields analysis rates all sectors as "out of normal limits"
- Temporal peripapillary atrophy (3)
- Decreased retinal reflectivity (4)
- Flattened and asymmetric height profile of the contour line (5) with depression in the inferior sectors (6)
- Both polar peaks ("double humps") (6+7) do not reach mean retina height (*)
- Corresponding visual field defect (MD 13.3 dB) (d)

a Profile of contour line, OS
b Intensitiy image, OS
c Stereometric parameters
d 30°-visual field, OS
e Topography image, OS

Plate 35

Comments

- Small optic disc (1)
- Very thin neuroretinal rim in the temporal/temporal inferior sector (2)
- Pathologically decreased global "rim volume" and "rim area"
- Pathologically increased "cup/disc area ratio" and "linear cup/disc ratio"
- Pathologically decreased "height variation contour" value
- Pathologic "cup shape measure" value
- Both discriminat functions rate optic disc as "glaucomatous"
- Moorfields analysis rates all sectors, except the temporal one, as "out of normal limits"
- Decreased retinal reflectivity (3)
- Flattened and asymmetric height profile of the contour line (4) with depression in the inferior sectors (5)
- Both polar peaks ("double humps") (5+6) do not reach mean retina height
- Corresponding visual field defect (MD 14.6 dB) (d)

a Profile of contour line, OD
b Intensitiy image, OD
c Stereometric parameters
d 30°-visual field, OD
e Topography image, OD

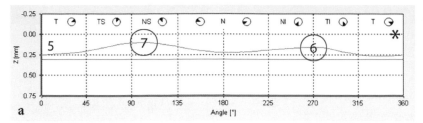

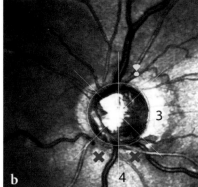

Parameters	global
disc area [mm²]	1.616
cup area [mm²]	1.041
rim area [mm²]	0.575
cup/disc area ratio [ ]	0.644
rim/disc area ratio [ ]	0.356
cup volume [mm³]	0.258
rim volume [mm³]	0.078
mean cup depth [mm]	0.286
maximum cup depth [mm]	0.554
height variation contour [mm]	0.167
cup shape measure [ ]	-0.026
mean RNFL thickness [mm]	0.116
RNFL cross sectional area [mm²]	0.523
linear cup/disc ratio [ ]	0.803
maximum contour elevation [mm]	0.101
maximum contour depression [mm]	0.268
CLM temporal-superior [mm]	0.098
CLM temporal-inferior [mm]	0.054
average variability (SD) [mm]	0.034
reference height [mm]	0.310
FSM discriminant function value [ ]	-2.208
RB discriminant function value [ ]	-0.839

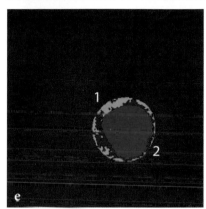

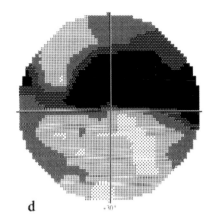

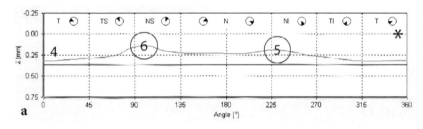

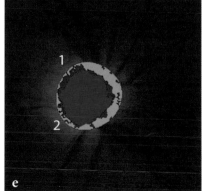

Parameters	global
disc area [mm²]	1.425
cup area [mm²]	0.830
rim area [mm²]	0.594
cup/disc area ratio [ ]	0.583
rim/disc area ratio [ ]	0.417
cup volume [mm³]	0.281
rim volume [mm³]	0.084
mean cup depth [mm]	0.353
maximum cup depth [mm]	0.686
height variation contour [mm]	0.184
cup shape measure [ ]	-0.019
mean RNFL thickness [mm]	0.115
RNFL cross sectional area [mm²]	0.487
linear cup/disc ratio [ ]	0.764
maximum contour elevation [mm]	0.141
maximum contour depression [mm]	0.326
CLM temporal-superior [mm]	0.075
CLM temporal-inferior [mm]	0.018
average variability (SD) [mm]	0.016
reference height [mm]	0.369
FSM discriminant function value [ ]	-2.323
RB discriminant function value [ ]	-0.560

Plate 36

Comments

- Small and tilted optic disc (1)
- Very thin neuroretinal rim in the temporal inferior sector (2)
- Decreased global "rim volume" and "rim area"
- Normal "cup/disc area ratio" and "linear cup/disc ratio"
- Increased "height variation contour" value
- Normal "cup shape measure" value
- FSM discriminat function rates optic disc as "glaucomatous", RB as "normal"
- Moorfields analysis rates only the temporal inferior sector as "out of normal limits"
- Temporal inferior peripapillary atrophy (3)
- Very asymmetric height profile of the contour line (4) with severe depression in the temporal inferior sector (5)
- Inferior segment polar peak (5) does not reach mean retina height ($*$) and even lies below the reference plane (red line)
- Corresponding visual field defect (d)

a Profile of contour line, OS
b Intensitiy image, OS
c Stereometric parameters
d 30°-visual field, OS
e Topography image OS

Plate 37

Comments

- Small optic disc (1)
- Very thin neuroretinal rim in the temporal sector (2)
- Pathologically decreased global "rim volume" and "rim area"
- Pathologically increased "cup/disc area ratio" and "linear cup/disc ratio"
- Pathologically decreased "height variation contour" value
- Pathologic "cup shape measure" value
- Both discriminat functions rate optic disc as "glaucomatous"
- Moorfields analysis rates all sectors except the nasal superior one as "out of normal limits"
- Temporal peripapillary atrophy (3)
- Decreased retinal reflectivity (4)
- Flattened and asymmetric height profile of the contour line (5) with depression in the superior sectors (6)
- Both polar peaks ("double humps") (6+7) do not reach mean retina height ($*$)
- Very advanced visual field defect (MD 20.1 dB) (d)

a Profile of contour line, OD
b Intensitiy image, OD
c Stereometric parameters
d 30°-visual field, OD
e Topography image, OD

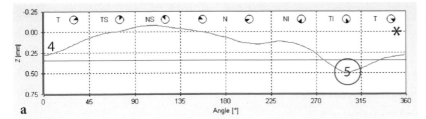

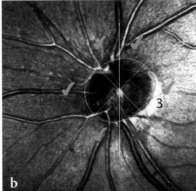

Parameters	global
disc area [mm²]	1.490
cup area [mm²]	0.342
rim area [mm²]	1.148
cup/disc area ratio [ ]	0.230
rim/disc area ratio [ ]	0.770
cup volume [mm³]	0.045
rim volume [mm³]	0.284
mean cup depth [mm]	0.151
maximum cup depth [mm]	0.485
height variation contour [mm]	0.581
cup shape measure [ ]	-0.258
mean RNFL thickness [mm]	0.194
RNFL cross sectional area [mm²]	0.840
linear cup/disc ratio [ ]	0.479
maximum contour elevation [mm]	-0.080
maximum contour depression [mm]	0.501
CLM temporal-superior [mm]	0.259
CLM temporal-inferior [mm]	-0.159
average variability (SD) [mm]	0.015
reference height [mm]	0.343
FSM discriminant function value [ ]	-0.110
RB discriminant function value [ ]	0.436

c

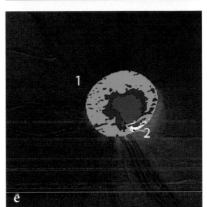

b

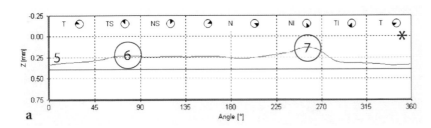

d   -30°

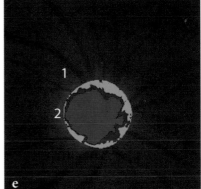

e

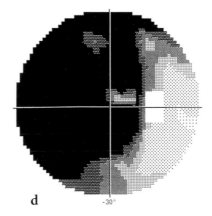

Parameters	global
disc area [mm²]	1.857
cup area [mm²]	1.078
rim area [mm²]	0.779
cup/disc area ratio [ ]	0.580
rim/disc area ratio [ ]	0.420
cup volume [mm³]	0.304
rim volume [mm³]	0.105
mean cup depth [mm]	0.313
maximum cup depth [mm]	0.754
height variation contour [mm]	0.217
cup shape measure [ ]	-0.091
mean RNFL thickness [mm]	0.131
RNFL cross sectional area [mm²]	0.634
linear cup/disc ratio [ ]	0.762
maximum contour elevation [mm]	0.131
maximum contour depression [mm]	0.348
CLM temporal-superior [mm]	0.079
CLM temporal-inferior [mm]	0.034
average variability (SD) [mm]	0.016
reference height [mm]	0.394
FSM discriminant function value [ ]	-1.631
RB discriminant function value [ ]	-0.401

c

d   -30°

e

Comments

- Very large optic disc (1)
- Nearly complete loss of the neuroretinal rim (2)
- "Step" in the temporal peripapillary retina (3) better visible in the 3D reconstruction (c)
- Pathologically decreased global "rim area" and "rim volume"
- Pathologically increased "cup/disc area ratio" and "linear cup/disc ratio"
- Very large circular cup (4)
- Pathologically decreased "height variation contour" value
- Pathologic "cup shape measure" value
- Low standard reference plane height
- Moorfields analysis rates all sectors "out of normal limits"
- Severly decreased retinal reflectivity (5)
- Flattened and asymmetric height profile of the contour line (6) with depression in the superior sectors (7)
- Both polar peaks ("double humps") (7+8) do not reach mean retina height (*)

Original printout with:
a Topography image, OD
b Intensity image, OD
c 3D-image, OD

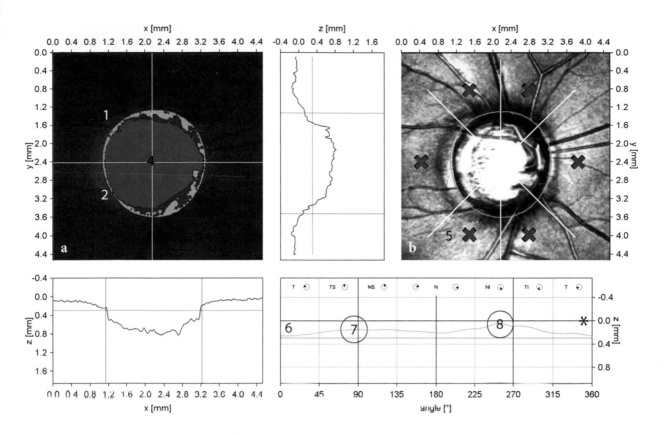

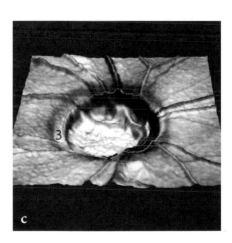

Stereometric Analysis ONH	
Disk Area	3.696 mm²
Cup Area	2.815 mm²
Rim Area	0.881 mm²
Cup Volume	0.857 cmm
Rim Volume	0.109 cmm
Cup/Disk Area Ratio	0.762
Linear Cup/Disk Ratio	0.873
Mean Cup Depth	0.366 mm
Maximum Cup Depth	0.617 mm
Cup Shape Measure	0.103
Height Variation Contour	0.214 mm
Mean RNFL Thickness	0.122 mm
RNFL Cross Sectional Area	0.834 mm²
Reference Height	0.294 mm
Topography Std Dev.	16 µm

Comments

- Very large optic disc (1)
- Circular thinned neuroretinal rim (2)
- Pathologically decreased global "rim area" and "rim volume"
- Pathologically increased "cup/disc area ratio" and "linear cup/disc ratio"
- Very large circular cup (3)
- Pathologically decreased "height variation contour" value
- Pathologic "cup shape measure" value
- Very low standard reference plane height (red line)
- Moorfields analysis rates all sectors as "out of normal limits"
- Severly decreased retinal reflectivity (4)
- Flattened and asymmetric height profile of the contour line (5) with depression in the inferior sectors (6)
- Inferior segment polar peak (6) does not reach mean retina height (*)

Original printout with:

a Topography image, OD
b Intensity image, OD
c Photograph of the optic disc, OD

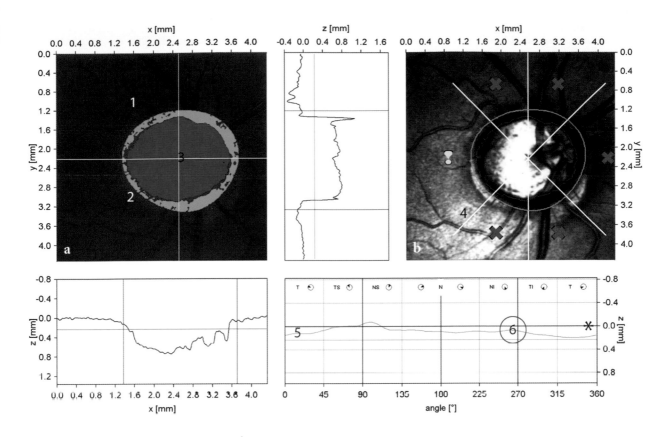

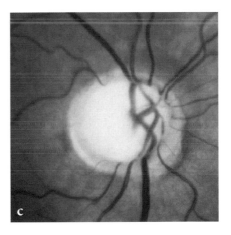

Stereometric Analysis ONH

Disk Area	3.818 mm²
Cup Area	2.669 mm²
Rim Area	1.148 mm²
Cup Volume	0.982 cmm
Rim Volume	0.213 cmm
Cup/Disk Area Ratio	0.699
Linear Cup/Disk Ratio	0.836
Mean Cup Depth	0.455 mm
Maximum Cup Depth	0.744 mm
Cup Shape Measure	0.087
Height Variation Contour	0.274 mm
Mean RNFL Thickness	0.139 mm
RNFL Cross Sectional Area	0.967 mm²
Reference Height	0.230 mm
Topography Std Dev.	15 µm

Comments

- Very large optic disc (1)
- Slightly decreased global "rim area" and "rim volume"
- Increased "cup/disc area ratio" and "linear cup/disc ratio"
- Large circular cup (2)
- Decreased "height variation contour" value
- Pathologic "cup shape measure" value
- "Step" in the temporal peripapillary retina (3)
- Moorfields analysis rates only the nasal/nasal superior sector as "out of normal limits"
- Flattened and asymmetric height profile of the contour line (4) with depression in the superior sectors (5)
- Superior segment polar peak (5) does not reach mean retina height (*)

Original printout with:
a Topography image, OD
b Intensitiy image, OD
c 30°-visual field, OD

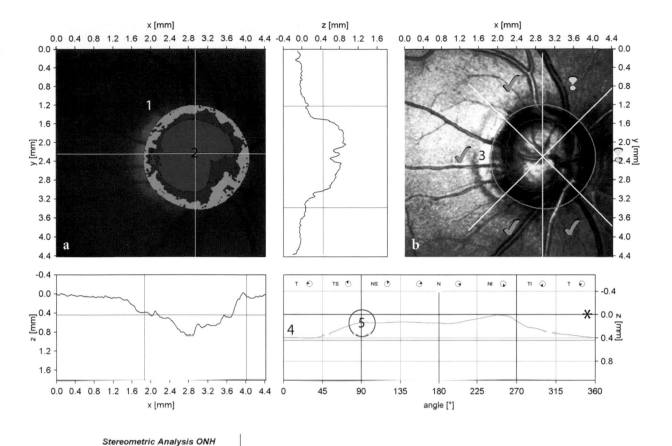

Stereometric Analysis ONH	
Disk Area	3.681 mm²
Cup Area	1.700 mm²
Rim Area	1.981 mm²
Cup Volume	0.376 cmm
Rim Volume	0.426 cmm
Cup/Disk Area Ratio	0.462
Linear Cup/Disk Ratio	0.680
Mean Cup Depth	0.307 mm
Maximum Cup Depth	0.682 mm
Cup Shape Measure	-0.077
Height Variation Contour	0.399 mm
Mean RNFL Thickness	0.230 mm
RNFL Cross Sectional Area	1.567 mm²
Reference Height	0.437 mm
Topography Std Dev.	10 µm

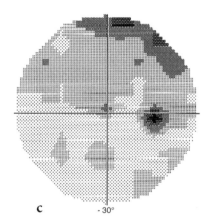

c   - 30°

Plate 41

Comments

- Very large optic disc (1)
- Normal global "rim area" and "rim volume"
- Increased "cup/disc area ratio" and "linear cup/disc ratio"
- Large circular cup with deep excavation (2)
- Normal "height variation contour" value
- Pathologic "cup shape measure" value
- Very high standard reference plane height (red line)
- Both discriminant functions rate optic disc as "normal"
- Temporal peripapillary atrophy (3)
- Moorfields analysis rates all nasal sectors and the temporal inferior one as "out of normal limits"
- Flattened height profile of the contour line (4)
- Both polar peaks ("double humps") (5) do not reach mean retina height (∗)

a Profile of contour line, OS
b Intensitiy image, OS
c Stereometric parameters
d 30°-visual field, OS
e Topography image OS

Plate 42

Comments

- Large optic disc (1)
- Pathologically decreased global "rim area" and "rim volume"
- Pathologically increased "cup/disc area ratio" and "linear cup/disc ratio"
- Very large circular cup with very deep excavation (2)
- Pathologically decreased "height variation contour" value
- Pathologic "cup shape measure" value
- Low standard reference plane height (red line)
- Both discriminant functions rate optic disc as "glaucomatous"
- Moorfields analysis rates all sectors as "out of normal limits"
- Decreased retinal reflectivity (3)
- Flattened height profile of the contour line (4)
- Both polar peaks ("double humps") (5) do not reach mean retina height (∗)

a Profile of contour line, OD
b Intensitiy image, OD
c Stereometric parameters
d 30°-visual field, OD
e Topography image, OD

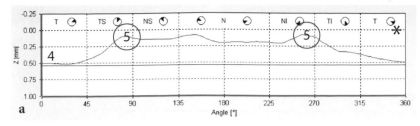

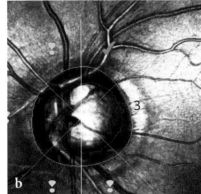

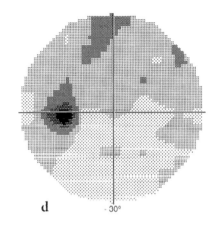

Parameters	global
disc area [mm²]	3.960
cup area [mm²]	1.913
rim area [mm²]	2.047
cup/disc area ratio [ ]	0.483
rim/disc area ratio [ ]	0.517
cup volume [mm³]	0.560
rim volume [mm³]	0.528
mean cup depth [mm]	0.387
maximum cup depth [mm]	0.832
height variation contour [mm]	0.439
cup shape measure [ ]	-0.062
mean RNFL thickness [mm]	0.277
RNFL cross sectional area [mm²]	1.957
linear cup/disc ratio [ ]	0.695
maximum contour elevation [mm]	0.079
maximum contour depression [mm]	0.518
CLM temporal-superior [mm]	0.240
CLM temporal-inferior [mm]	0.194
average variability (SD) [mm]	0.020
reference height [mm]	0.538
FSM discriminant function value [ ]	0.784
RB discriminant function value [ ]	0.881

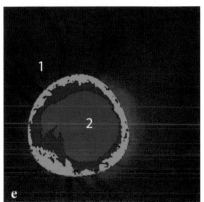

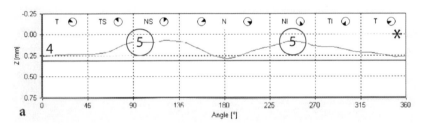

Parameters	global
disc area [mm²]	3.117
cup area [mm²]	2.226
rim area [mm²]	0.891
cup/disc area ratio [ ]	0.714
rim/disc area ratio [ ]	0.286
cup volume [mm³]	1.251
rim volume [mm³]	0.148
mean cup depth [mm]	0.618
maximum cup depth [mm]	1.079
height variation contour [mm]	0.216
cup shape measure [ ]	0.034
mean RNFL thickness [mm]	0.134
RNFL cross sectional area [mm²]	0.842
linear cup/disc ratio [ ]	0.845
maximum contour elevation [mm]	0.074
maximum contour depression [mm]	0.290
CLM temporal-superior [mm]	0.067
CLM temporal-inferior [mm]	0.074
average variability (SD) [mm]	0.035
reference height [mm]	0.316
FSM discriminant function value [ ]	-3.451
RB discriminant function value [ ]	-1.050

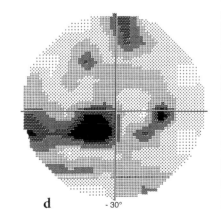

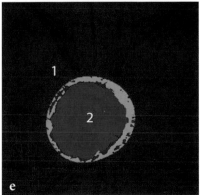

Comments

- Large optic disc (1)
- Vessel trunk simulates a normal neuroretinal rim in the nasal/nasal inferior sector (2)
- Decreased global "rim area" and "rim volume"
- Pathologically increased "cup/disc area ratio" and "linear cup/disc ratio"
- Vertically pronounced cup with shallow excavation (3)
- Pathologically decreased "height variation contour" value
- Pathologic "cup shape measure" value
- Moorfields analysis rates only the two superior sectors as "out of normal limits"
- Decreased retinal reflectivity (4)
- Flattened height profile of the contour line (5) with depression in the superior sectors (6)
- Both polar peaks ("double humps") (6+7) do not reach mean retina height (∗)

Original printout with:

a Topography image, OS
b Intensity image, OS
c Photograph of the optic disc, OS

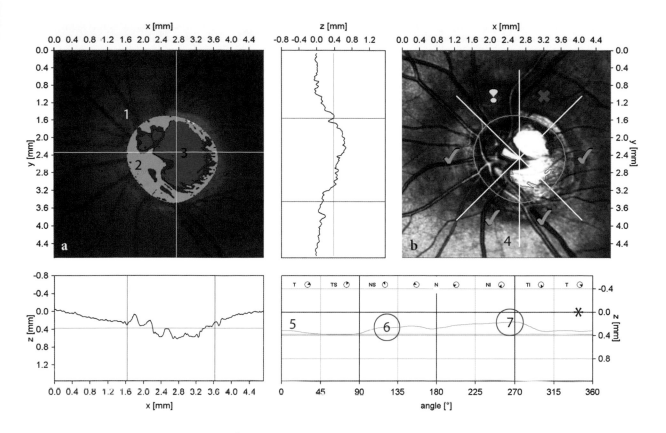

Stereometric Analysis ONH	
Disk Area	2.977 mm²
Cup Area	1.264 mm²
Rim Area	1.713 mm²
Cup Volume	0.146 cmm
Rim Volume	0.234 cmm
Cup/Disk Area Ratio	0.424
Linear Cup/Disk Ratio	0.652
Mean Cup Depth	0.145 mm
Maximum Cup Depth	0.331 mm
Cup Shape Measure	-0.078
Height Variation Contour	0.221 mm
Mean RNFL Thickness	0.092 mm
RNFL Cross Sectional Area	0.564 mm²
Reference Height	0.377 mm
Topography Std Dev.	12 µm

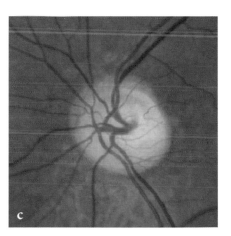

Comments

- Circular thinned neuroretinal rim (1)
- Pathologically decreased global "rim area" and "rim volume"
- Pathologically increased "cup/disc area ratio" and "linear cup/disc ratio"
- Vertically pronounced cup with shallow excavation (2)
- Pathologically decreased "height variation contour" value
- Pathologic "cup shape measure" value
- Very low standard reference plane height (red line)
- Moorfields analysis rates all sectors except the temporal and temporal superior one as "out of normal limits"
- Circular peripapillary atrophy (3)
- Decreased retinal reflectivity (4)
- Flattened height profile of the contour line (5)
- Both polar peaks ("double humps") (6) do not reach mean retina height (*)

Original printout with:
a Topography image, OD
b Intensitiy image, OD
c 30°-visual field, OD

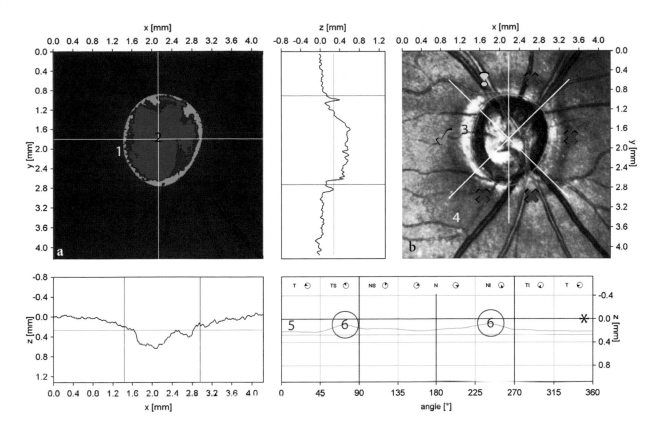

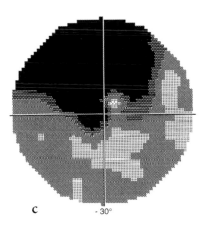

Stereometric Analysis ONH	
Disk Area	2.186 mm²
Cup Area	1.419 mm²
Rim Area	0.767 mm²
Cup Volume	0.249 cmm
Rim Volume	0.068 cmm
Cup/Disk Area Ratio	0.649
Linear Cup/Disk Ratio	0.806
Mean Cup Depth	0.213 mm
Maximum Cup Depth	0.468 mm
Cup Shape Measure	0.047
Height Variation Contour	0.142 mm
Mean RNFL Thickness	0.090 mm
RNFL Cross Sectional Area	0.475 mm²
Reference Height	0.267 mm
Topography Std Dev.	49 µm

Plate 45

Comments

- Large optic disc (1)
- Nearly complete loss of the neuroretinal rim (2)
- Pathologically decreased global "rim area" and "rim volume"
- Pathologically increased "cup/disc area ratio"
- Very large cup (3)
- Pathologically decreased "height variation contour" value
- Pathologic "cup shape measure" value
- Very low standard reference plane height
- Circular peripapillary atrophy zone (4)
- Severly decreased retinal reflectivity (5)
- Severly flattened height profile of the contour line (6)
- Both polar peaks ("double humps") (7) do not reach mean retina height (∗)

Original printout with:
a Topography image, OD
b Intensitiy image, OD
c Photograph of the optic disc, OD

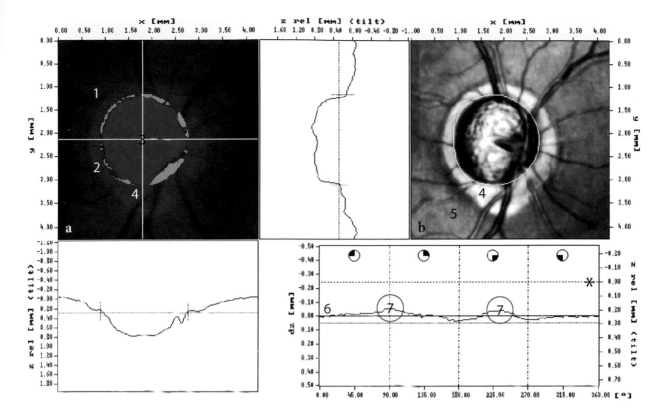

Stereometric Analysis ONH:

Disk Area:	2.950	mm²
Cup Area:	2.412	mm²
Cup/Disk Area Ratio:	0.818	
Rim Area:	0.537	mm²
Cup Volume:	0.873	cmm
Rim Volume:	0.040	cmm
Mean Cup Depth:	0.377	mm
Maximum Cup Depth:	0.623	mm
Cup Shape Measure:	0.103	
Height Variation Contour:	0.009	mm
Mean RNFL-Thickness:	0.054	mm
RNFL-Cross Section Area:	0.331	mm²
Reference Height (Std.):	0.293	mm

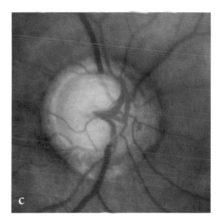

Plate 46

Comments

- Medium-sized optic disc (1)
- Nearly complete loss of the neuroretinal rim (2)
- Pathologically decreased global "rim area" and "rim volume"
- Pathologically increased "cup/disc area ratio"
- Large cup with shallow excavation (3)
- Pathologically decreased "height variation contour" value
- Pathologic "cup shape measure" value
- Very low standard reference plane height
- Circular peripapillary atrophy zone (4)
- Severly decreased retinal reflectivity (5)
- Severely flattened height profile of the contour line (6)
- Both polar peaks ("double humps") (7) do not reach mean retina height (*)

Original printout with:
a Topography image, OS
b Intensitiy image, OS
c Photograph of the optic disc, OS

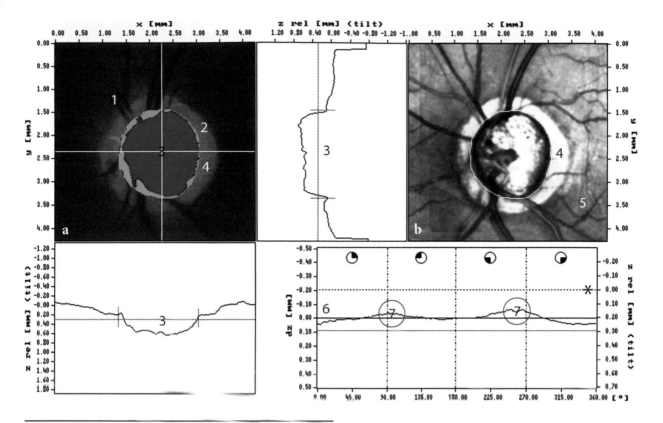

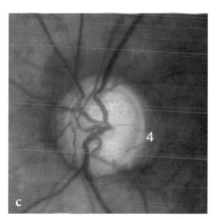

Stereometric Analysis ONH:

Disk Area:	2.660	mm²
Cup Area:	2.094	mm²
Cup/Disk Area Ratio:	0.787	
Rim Area:	0.566	mm²
Cup Volume:	0.512	cmm
Rim Volume:	0.068	cmm
Mean Cup Depth:	0.311	mm
Maximum Cup Depth:	0.505	mm
Cup Shape Measure:	0.109	
Height Variation Contour:	0.112	mm
Mean RNFL-Thickness:	0.095	mm
RNFL-Cross Section Area:	0.549	mm²
Reference Height (Std.):	0.295	mm

Plate 47

Comments

- Very thin neuroretinal rim in temporal/temporal inferior sector (1)
- Vessel trunk simulates normal neuroretinal rim in the nasal/nasal superior sector (2)
- Pathologically decreased global "rim area" and "rim volume"
- Pathologically increased "cup/disc area ratio" and "linear cup/disc area ratio"
- Vertically pronounced cup shape (3)
- Pathologically decreased "height variation contour" value
- Pathologic "cup shape measure" value
- Moorfields analysis rates all sectors except the nasal and nasal superior one as "out of normal limits"
- Circular peripapillary atrophy zone (4)
- Severly decreased retinal reflectivity (5)
- Flattened and asymmetric height profile of the contour line (6) with depression in the inferior sectors (7)
- Both polar peaks ("double humps") (7+8) do not reach mean retina height (*)

a Profile of contour line, OS
b Intensitiy image, OS
c Stereometric parameters
d 30°-visual field, OS
e Topography image OS

Plate 48

Comments

- Very thin neuroretinal rim in temporal sectors and nasal superior sector (1)
- Vessel trunk simulates a normal neuroretinal rim in the nasal inferior sector (2)
- Pathologically decreased global "rim area" and "rim volume"
- Pathologically increased "cup/disc area ratio" and "linear cup/disc area ratio"
- Vertically pronounced cup with shallow excavation (3)
- Pathologically decreased "height variation contour" value
- Pathologic "cup shape measure" value
- Moorfields analysis rates all sectors except the temporal one as "out of normal limits"
- Circular peripapillary atrophy zone (4)
- Severly decreased retinal reflectivity (5)
- Flattened height profile of the contour line (6)
- Both polar peaks ("double humps") (7) do not reach mean retina height (*)

a Profile of contour line, OD
b Intensitiy image, OD
c Stereometric parameters
d 30°-visual field, OD
e Topography image, OD

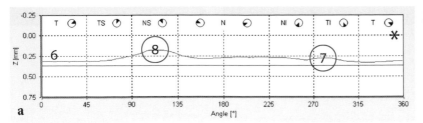

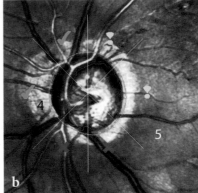

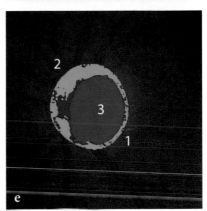

Parameters	global
disc area [mm²]	2.020
cup area [mm²]	1.179
rim area [mm²]	0.841
cup/disc area ratio [ ]	0.584
rim/disc area ratio [ ]	0.416
cup volume [mm³]	0.310
rim volume [mm³]	0.098
mean cup depth [mm]	0.283
maximum cup depth [mm]	0.530
height variation contour [mm]	0.155
cup shape measure [ ]	-0.008
mean RNFL thickness [mm]	0.083
RNFL cross sectional area [mm²]	0.421
linear cup/disc ratio [ ]	0.764
maximum contour elevation [mm]	0.179
maximum contour depression [mm]	0.334
CLM temporal-superior [mm]	0.028
CLM temporal-inferior [mm]	0.025
average variability (SD) [mm]	0.025
reference height [mm]	0.365
FSM discriminant function value [ ]	-2.024
RB discriminant function value [ ]	-1.306

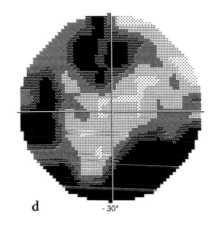

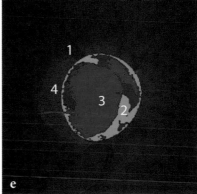

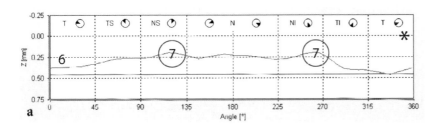

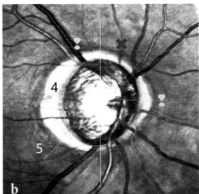

Parameters	global
disc area [mm²]	2.755
cup area [mm²]	1.688
rim area [mm²]	1.068
cup/disc area ratio [ ]	0.613
rim/disc area ratio [ ]	0.387
cup volume [mm³]	0.216
rim volume [mm³]	0.165
mean cup depth [mm]	0.227
maximum cup depth [mm]	0.473
height variation contour [mm]	0.263
cup shape measure [ ]	-0.019
mean RNFL thickness [mm]	0.156
RNFL cross sectional area [mm²]	0.921
linear cup/disc ratio [ ]	0.783
maximum contour elevation [mm]	0.200
maximum contour depression [mm]	0.463
CLM temporal-superior [mm]	0.116
CLM temporal-inferior [mm]	0.042
average variability (SD) [mm]	0.027
reference height [mm]	0.455
FSM discriminant function value [ ]	-2.187
RB discriminant function value [ ]	-0.477

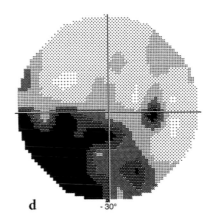

Comments

- Relatively small optic disc (1)
- Very thin neuroretinal rim in the temporal sector (2)
- Vessel trunk simulates a normal neuroretinal rim in the nasal hemisphere of the optic disc (3)
- Slightly decreased global "rim area" and "rim volume"
- Slightly increased "cup/disc area ratio" and "linear cup/disc area ratio"
- Vertically pronounced cup shape (4)
- Normal "height variation contour" value
- Pathologic "cup shape measure" value
- Moorfields analysis rates only the temporal sector as "out of normal limits"
- Temporal peripapillary atrophy zone (5)
- Decreased retinal reflectivity (6)
- Flattened and asymmetric height profile of the contour line (7) with depression in the temporal inferior sectors (8)
- Inferior segment polar peak (8) does not reach mean retina height (*)
- Advanced visual field defect (MD 9.7 dB) corresponds to damage in inferior nerve fiber bundels (c)

Original printout with:
a Topography image, OS
b Intensitiy image, OS
c 30°-visual field, OS

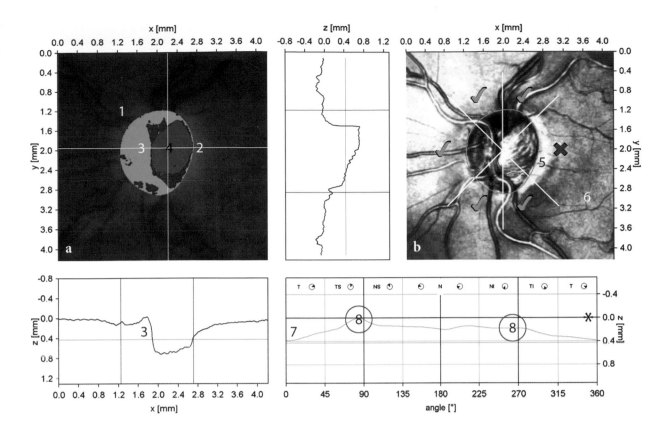

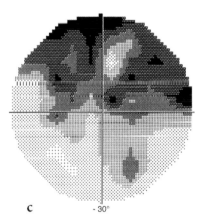

Stereometric Analysis ONH	
Disk Area	1.984 mm²
Cup Area	0.684 mm²
Rim Area	1.300 mm²
Cup Volume	0.116 cmm
Rim Volume	0.389 cmm
Cup/Disk Area Ratio	0.345
Linear Cup/Disk Ratio	0.587
Mean Cup Depth	0.262 mm
Maximum Cup Depth	0.679 mm
Cup Shape Measure	-0.057
Height Variation Contour	0.384 mm
Mean RNFL Thickness	0.222 mm
RNFL Cross Sectional Area	1.109 mm²
Reference Height	0.425 mm
Topography Std Dev.	14 μm

Plate 50

Comments

- Very thin neuroretinal rim in the temporal sector (1)
- Vessel trunk simulates normal neuroretinal rim in the nasal hemisphere of the optic disc (2)
- Slightly decreased global "rim area" and "rim volume"
- Increased "cup/disc area ratio" and "linear cup/disc area ratio"
- Vertically pronounced cup shape (3)
- Decreased "height variation contour" value
- Borderline "cup shape measure" value
- Increased standard reference height (red line)
- Discriminant functions rate optic disc as "normal"
- Moorfields analysis rates only the temporal and the temporal inferior sector as "out of normal limits"
- Decreased retinal reflectivity (4)
- Flattened and asymmetric height profile of the contour line (5) with depression in the inferior sectors (6)
- Both polar peaks ("double humps") (6+7) do not reach mean retina height (∗)
- Advanced visual field defect (MD 10.1 dB) corresponds to damage in inferior nerve fiber bundels

a Profile of contour line, OS
b Intensitiy image, OS
c Stereometric parameters
d 30°-visual field, OS
e Topography image OS

Plate 51

Comments

- Slightly tilted optic disc
- Vessel trunk simulates normal neuroretinal rim in the nasal hemisphere of the optic disc (1)
- Moorfields analysis rates only the temporal sector as "out of normal limits" (not shown here)
- Decreased retinal reflectivity (2)
- Severely flattened amplitudes of visual evoked potentials of the right eye (3) demonstrate optic atrophy
- Advanced visual field defects (c) secondary to non-glaucomatous optic atrophy of the right eye

Case: A 63-year-old male with optic atrophy of traumatic origin (car accident)

a Visual evoced potentials
b Intensity image of interactive measuremet, OD
c 30°-visual field, OD

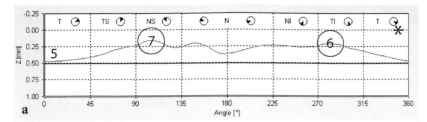

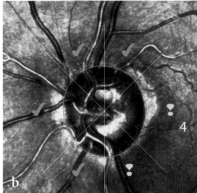

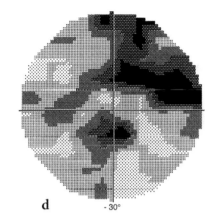

Parameters	global
disc area [mm²]	2.705
cup area [mm²]	1.188
rim area [mm²]	1.517
cup/disc area ratio [ ]	0.439
rim/disc area ratio [ ]	0.561
cup volume [mm³]	0.224
rim volume [mm³]	0.321
mean cup depth [mm]	0.252
maximum cup depth [mm]	0.650
height variation contour [mm]	0.315
cup shape measure [ ]	-0.117
mean RNFL thickness [mm]	0.202
RNFL cross sectional area [mm²]	1.179
linear cup/disc ratio [ ]	0.663
maximum contour elevation [mm]	0.161
maximum contour depression [mm]	0.476
CLM temporal-superior [mm]	0.117
CLM temporal-inferior [mm]	0.182
average variability (SD) [mm]	0.031
reference height [mm]	0.509
FSM discriminant function value [ ]	0.808
RB discriminant function value [ ]	0.629

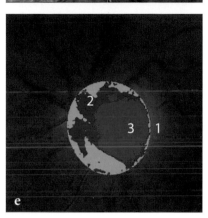

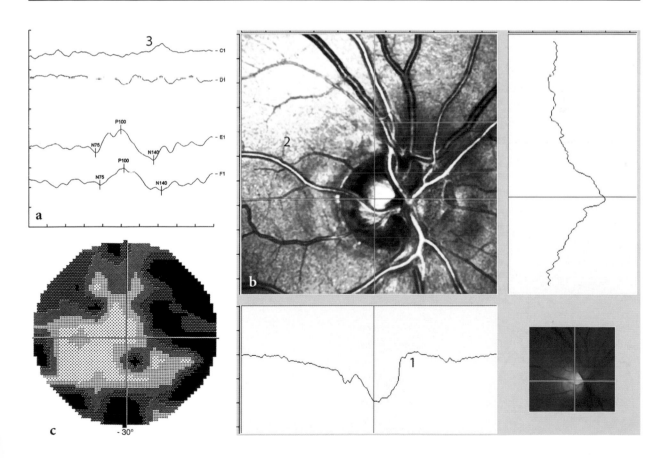

Comments

- Very thin neuroretinal rim in the temporal sector (1)
- Vessel trunk simulates normal neuroretinal rim (2) in the nasal hemisphere of the optic disc
- Slightly decreased global "rim area" and "rim volume"
- Increased "cup/disc area ratio" and "linear cup/disc area ratio"
- Vertically pronounced cup shape (3)
- Decreased "height variation contour" value
- Borderline "cup shape measure" value
- Moorfields analysis rates only the temporal and the temporal superior sector as "out of normal limits"
- Decreased retinal reflectivity (4)
- Flattened and asymmetric height profile of the contour line (5)
- Both polar peaks ("double humps") (6) do not reach mean retina height (*)
- Fundusphotograph (c) shows temporal atrophy of the right optic nerve head
- Nasal hemispheric visual field defect (d) corresponding to intracranial damage

Case: A 64-year-old male with non-glaucomatous optic atrophy of the right eye after clipping of an intracranial aneurysm two months prior to examination.

Original printout with:
a Topography image, OD
b Intensitiy image, OD
c Photograph of the optic disc, OD
d 30°-visual field, OD

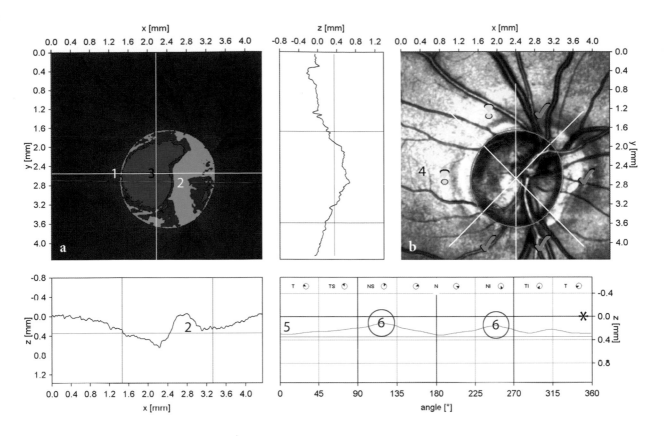

Stereometric Analysis ONH

Disk Area	2.950 mm²
Cup Area	1.150 mm²
Rim Area	1.800 mm²
Cup Volume	0.123 cmm
Rim Volume	0.257 cmm
Cup/Disk Area Ratio	0.390
Linear Cup/Disk Ratio	0.624
Mean Cup Depth	0.138 mm
Maximum Cup Depth	0.360 mm
Cup Shape Measure	-0.141
Height Variation Contour	0.205 mm
Mean RNFL Thickness	0.106 mm
RNFL Cross Sectional Area	0.643 mm²
Reference Height	0.341 mm
Topography Std Dev.	16 µm

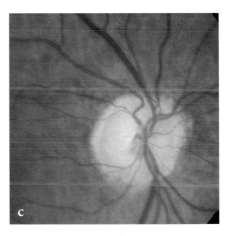

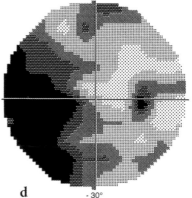

4 Follow-up Examinations

A. F. SCHEUERLE, E. SCHMIDT

4.1 Strategies for Longitudinal Analysis

Precise and objective follow-up analysis is definitely one of the key functions of laser scanning ophthalmoscopy. However, only a few prospective longitudinal studies from glaucoma patients using laser scanning tomography have been published to date. Initial studies, performed with the Laser Tomographic Scanner, pointed out that one must have a follow-up period of more than one year to detect optic disc changes in patients with glaucoma (Rohrschneider et al. 1994). Presumably, this is secondary to the appropriate therapy patients receive after diagnosis. Therefore, patients with ocular hypertension, lacking visual field defects, might be a suitable population for the study of optic disc changes. Of note, it has been shown that sequential analysis of laser scanning tomographies allows the detection of significant optic disc changes before confirmed visual field change in a group of ocular hypertensive patients converting to early glaucoma (Kamal et al. 1999).

Generally, there are two different methods for follow-up analysis of HRT topographies: on the one hand, changes in stereometric parameters between two examinations can be quantified; on the other hand, the difference of digital local height measurements of two examinations may be calculated. The first method is more suitable for quantitative evaluation of changes, while the second allows for a better localization of changes within the image. To compare two or more mean topographies with each other, the images have to be brought in the same perspective, rotation, tilt and magnification (normalization). HRT software, however, usually performs normalization automatically. Despite the elaborated correction mechanism, difficulties could arise if the angle or the distance between the eye and the scanner are too different during two sessions. Therefore, the examiner(s) should try to keep the head and the eyes in about the same standard position for any image acquisition.

To quantify the change of stereometric parameters, the baseline's contour line has to be used in the follow-up examinations after the normalization procedure. Now, differences between images due to different imaging conditions are almost eliminated and baseline and follow-up examinations may be compared. As the dimensions of the stereometric parameters calculated by HRT software are inhomogenius, an equitation that expresses the normalized change of values has been developed. The normalized change of any parameter is 0 if the parameter value is stable over time; the normalized change of any parameter is -1 if an average normal eye converts to advanced glaucoma. The normalized changes are dimensionless and can therefore be displayed on the same scale. Furthermore, it is possible to compute the average change of parameter clusters from a selected sector of the optic disc. The actual HRT software offers three different sector combinations (temporal superior and inferior octant, superior and inferior sector, and upper and lower hemisphere) that are always displayed together with the global normalized change.

In contrast to the change of stereometric parameters, the difference of the digital local height measurements (pixel) of two images can be calculated without any reference plane or contour line. After normalization of the mean topographies, the local height values from two different examinations are subtracted from each other. The resulting differences must be greater than the standard deviation of every single pixel and are displayed in a color-coded map. An overlap of standard deviations indicates signal noise and is therefore not displayed. Regions that are more depressed in the follow-up examination appear in red color while regions that are more elevated in the follow-up examination appear in green color superimposed on the reflectance image.

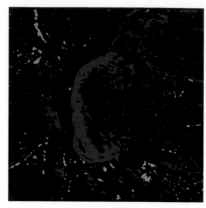

Fig. 11. Single pixel analysis of baseline and follow-up exam after 4 years

Detection of a statistically significant change in the same optic disc over time requires comparison of the change in an index with the variability in its repeated measurement. The significance of local height changes has been markedly enhanced by the introduction of a change probability map analysis developed by Chauhan et al., (Chauhan 1996, Chauhan et al. 2000). The 384×384 (HRT 256×256)-pixel array from each single topographic image is divided into a 96×96 (HRT 64×64)-superpixel array where each superpixel contains 16 (4×4) pixels. Usually, three series of images are obtained in each examination. The three corresponding topographies are calculated and normalized with the three topographies of the follow-up examination. Now, for each superpixel, there are 48 baseline and 48 follow-up height measurements. Accepting a lower spatial resolution, 96 height measurements in each superpixel (compared to only 6 height measurements in the single pixel analysis) can be used to perform an F test. An analysis of variance is conducted to detect changes in topographic heights from one set of images to the other. Superpixels with an error probability of less than 5% for rejecting the equal variances hypothesis indicate a significant change at the corresponding location. The analysis output is a probability map in which the probability that the difference in topographic height within a given superpixel occurs by chance alone is shown on a gray scale (unchanged=white, insignificant change=gray, significant change=black). A confirmed change in the topographic height within a given superpixel was considered to be present if the significance value associated with it was *always 5% or less in three consecutive sets of follow-up images.* Figures 11–13 illustrate improved display of significant local height changes in the original digital subtraction comparing only two exams and the current probability map analysis with significance values of less than 5% in three consecutive sets of follow-up images. In the case of a glaucoma patient, the change suspected after simple digital subtraction turned out to be insignificant using probability map analysis. A recent longitudinal study of patients with early glaucoma revealed that over a median follow-up of 5.5 years, disc changes determined by laser scanning tomography and change probability map analysis occur more frequently than field changes (Chauhan et al. 2001). These findings support the hypothesis that disc changes are more

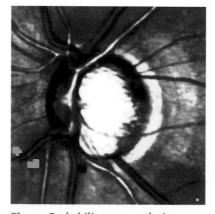

Fig. 12. Superpixel analysis of the same eye

Fig. 13. Probability map analysis of three consecutive sets of follow-up images effectively eliminates insignificant variation between single exams

sensitive to glaucomatous progression than visual field changes. However, it is to be noted that also this sophisticated method is purely a statistical description of the probability that the topographic height within a super-pixel has changed between different examinations. A statistically significant change that, for example, regularly occurs over vessels secondary to pulsations, might be irrelevant under clinical aspects.

The aim of regular sequential analysis of laser scanning tomographies is to detect optic disc changes not only in glaucoma patients but also in glaucoma suspects and to initiate therapy as soon as necessary, knowing that by the time visual field defects become evident, the eye has already been irreversibly damaged. The change probability map provides a high specificity in the analysis of disc changes but requires at least two follow-up exams. We therefore recommend more frequent imaging examinations with intervals between 3 and 12 months, depending on the individual patient's risk.

Plate 53

Comments

- The difference image of digital local height measurements of two examinations after three years (a) shows large confluating areas of red color (depression) in the temporal sectors of the disc and the temporal inferior peripapillary retina
- We conclude that this extrapapillary depression indicates a beginning nerve fiber bundle defect secondary to glaucomatous damage
- The temporal superior depression is limited to the optic disc and represents an area typically affected from glaucomatous progression
- The spreaded (not confluating) pieces of color in other areas of the difference image probably do not express clinically significant changes in topography
- However, in contrast to the results of HRT II, evaluating the difference images of the HRT topographic changes may be localized, but not quantified

Case: A 54-year-old female was diagnosed as having primary open-angle glaucoma. Local therapy was initiated but unfortunately the patient missed the scheduled appointments and returned to the first follow-up exam after three years. Besides the described topographic changes, static perimetry demonstrated new visual field defects.

Original HRT printout with:
a Topography difference image, OD
b Follow-up intensity image, OD

Plate 54

Comments

- The comparison between the two photographs (a, b) reveals increased vessel kinking in the inferior sector and a new intraretinal bleeding (b)
- The difference image of digital local height measurements of two examinations (c) shows two confluating areas of depression (red color) in the superior and inferior sector of the disc and in the temporal inferior peripapillary retina
- We conclude that this extrapapillary depression indicates a beginning nerve fiber bundle defect secondary to glaucomatous damage
- The superior and inferior depressions indicate loss of neuroretinal rim secondary to glaucomatous progression
- The spreaded (not confluating) pieces of green color are horizontally arranged and might result from small astigmatic changes

Case: Note the similar development of glaucomatous damage in the right (upper plate) and left optic disc of this patient. Also the left eye presented new visual field defects (f) after three years.

a Baseline photograph, OS
b Follow-up photograph, OS
c Topography difference image, OS
d Baseline intensity image, OS
e Follow-up intensity image, OS
f 30°-visual field

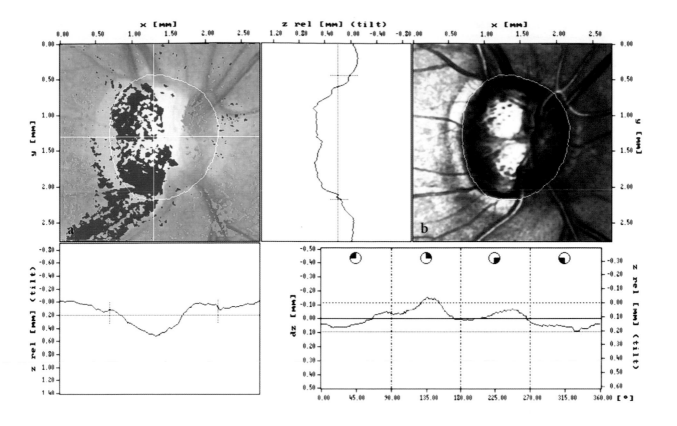

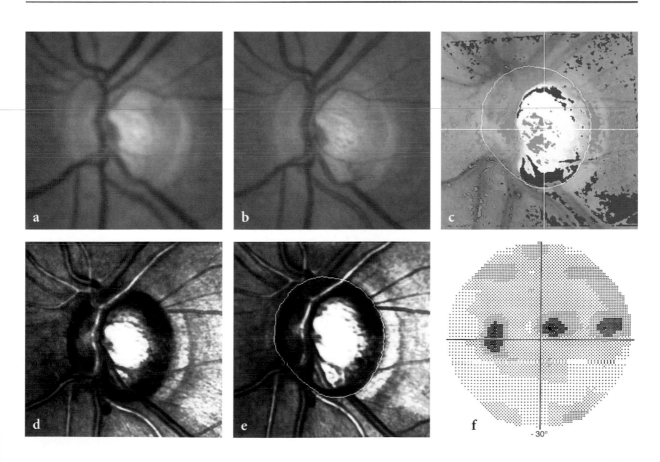

Plate 55

Comments

- The black and white image (a) indicates a follow-up exam. In the first follow-up image the HRT II software does not show color coded topographic differences of digital local height measurements
- The intensity image (b) shows a tilted disc which was suspicious for beginning glaucoma
- The stereometric progression chart (e) shows changes of normalized parameters. In this case the first follow-up indicates a small decrease of values after one year
- Three months later the second follow-up confirms the decrease and enables superpixel progression analysis (d)
- The probability map (d) shows a large confluating area of depression (red color) in the superior rim area
- We conclude that some portions of this area might be influenced by vessel pulsations while the changes of the disc border indicate loss of neuroretinal rim secondary to glaucomatous progression

Case: A 62-year-old female was under investigation for mild ocular hypertension. After fifteen months, topographic changes could be confirmed. Static perimetry (c) demonstrated new visual field defects and antiglaucomatous therapy was initiated.

Original HRT II printout with:
a Topography image b/w, OS
b Intensity image, OS
c 30°-visual field, OS
d Progression analysis, OS
e Stereometric progression chart

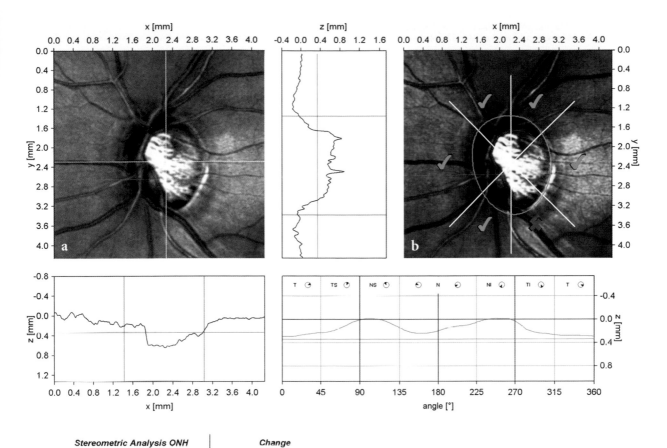

Stereometric Analysis ONH		Change
Disk Area	2.601	0.000 mm²
Cup Area	1.279	0.057 mm²
Rim Area	1.322	-0.057 mm²
Cup Volume	0.234	0.012 cmm
Rim Volume	0.267	-0.082 cmm
Cup/Disk Area Ratio	0.492	0.022
Linear Cup/Disk Ratio	0.701	0.015
Mean Cup Depth	0.255	-0.034 mm
Maximum Cup Depth	0.659	0.011 mm
Cup Shape Measure	-0.133	-0.056
Height Variation Contour	0.311	-0.022 mm
Mean RNFL Thickness	0.184	-0.032 mm
RNFL Cross Sectional Area	1.056	-0.181 mm²
Reference Height	0.336	-0.016 mm
Topography Std Dev.	12	µm

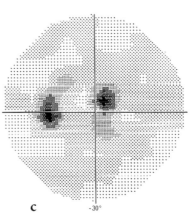

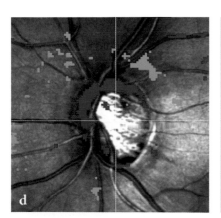

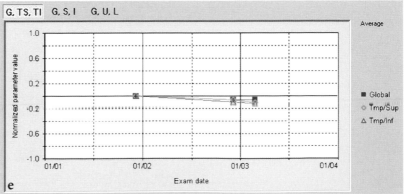

Plate 56

Comments

- The follow-up image (b) shows differences of digital local height measurements between the two examinations
- The progression analysis reveals a large confluating area of depression (red color) in the temporal superior rim area
- We conclude that this depression indicates topographic changes in an area that is often affected from glaucomatous progression

Case: A 65-year-old female with a history of primary open-angle glaucoma returning to scheduled control exams every six months.

Progression analysis with:
a Baseline intensity image, OD
b Follow-up intensity image, OD

Plate 57

Comments

- The baseline intensity image clearly demonstrates a nerve fiber bundle defect in the temporal inferior peripapillary retina
- The progression analysis (b) shows a large confluating area of depression (red color) along the temporal rim area
- We conclude that this depression indicates topographic changes in an area that may be affected from glaucomatous progression

Case: While the right (upper plate) and left disc appear different in shape the amount of visual field defects (MD 4–5 dB) was similar in both eyes.

Progression analysis with:
a Baseline intensity image, OS
b Follow-up intensity image, OS

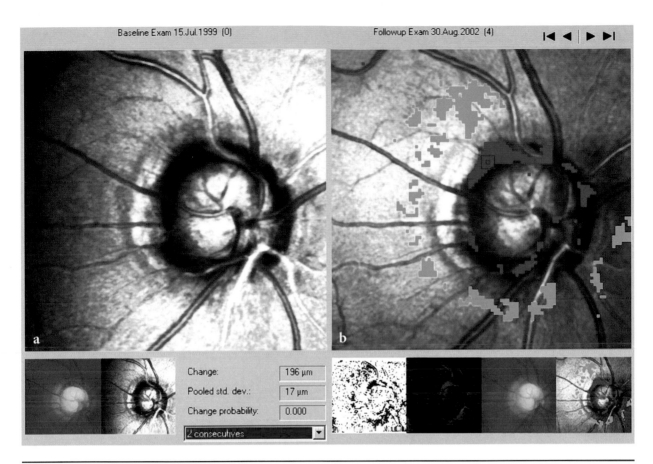

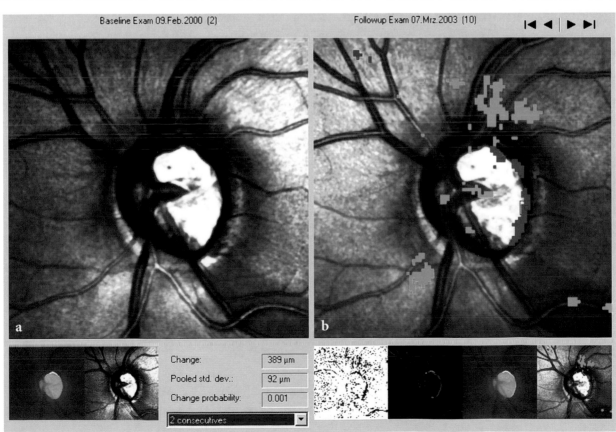

Plate 58

Comments

- The photograph (a) shows a small optic disc with a vertically pronounced excavation and notably thinned neuroretinal rim in the superior sectors
- The progression analysis (b) reveals a significant depression (red color) of circular shape sparing only the temporal sector
- We conclude that these topographic changes indicate a rapid and severe progression of glaucomatous damage
- The origin of the green areas remains unknown

Case: A 42-year-old male with a history of pseudoexfoliation glaucoma presented with high intraocular pressure not sufficiently controlled by local therapy and was therefore scheduled for trabeculotomy.

Note that the described topographic changes were observed over a period of only four months.

Progression analysis with:
a Follow-up photograph, OD
b Follow-up intensity image, OD

Plate 59

Comments

- The topography difference image (a) shows a circular depression (red color) of the neuroretinal rim
- We conclude that this depression indicates topographic changes secondary to glaucomatous progression
- Note the very steep and extremely deep excavation demonstrated by horizontal and vertical cross-sections

Case: A 68-year-old female with a long history of primary open-angle glaucoma, advanced visual filed defects and optic atrophy.

Original HRT printout with:
a Topography difference image, OD
b Follow-up intensity image, OD

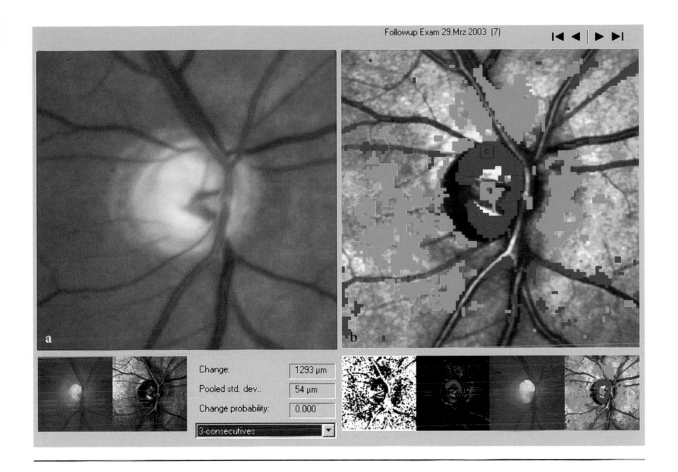

Change:	1293 µm
Pooled std. dev.:	54 µm
Change probability:	0.000

3-consecutives

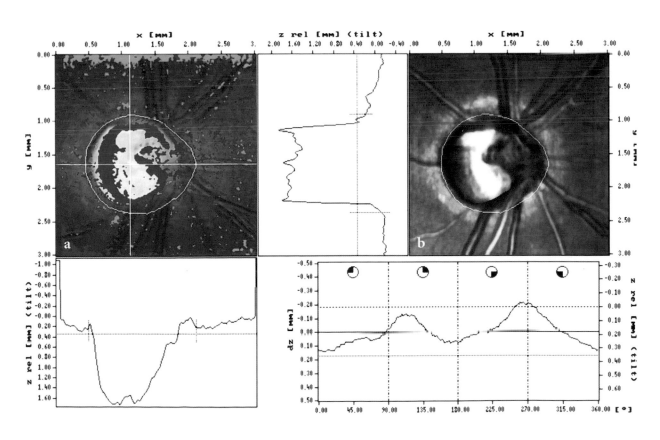

Plate 60

Comments

- The baseline intensity image (a) shows a relatively small optic disc with very large and shallow excavation characteristic for normal tension glaucoma
- The progression analysis (b) demonstrates a large confluating area of depression (red color) along the temporal disc border
- We conclude that this depression indicates significant topographic changes in a glaucomatous optic disc with preexisting advanced atrophy

Case: A 75-year-old female with a history of normal-tension glaucoma who missed the scheduled appointments and returned to the first follow-up exam after three years.

Progression analysis with:
a Baseline intensity image, OD
b Follow-up intensity image, OD

Plate 61

Comments

- The baseline intensity image (a) shows a macropapilla with large excavation and slightly reduced stereometric parameters
- The progression analysis demonstrates no significant changes over a follow-up period of nearly four years

Case: A 49-year-old female with primary open-angle glaucoma and beginning visual field defects (MD 5.7 dB). The initially elevated intraocular pressure was regulated with local combination therapy. Repeated visual field testing confirmed a stable state.

Progression analysis with:
a Baseline intensity image, OS
b Follow-up intensity image, OS

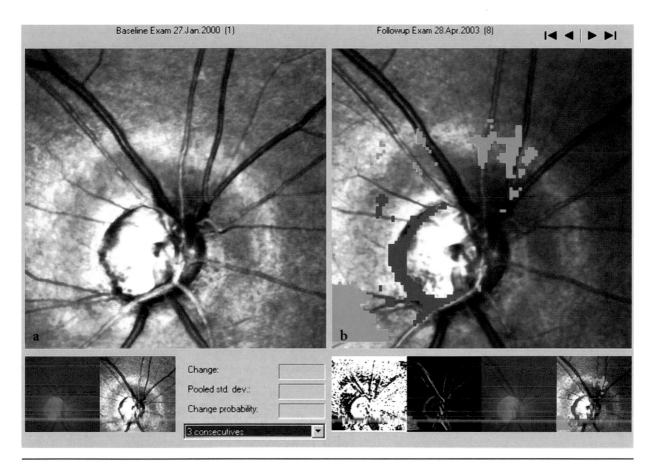

Baseline Exam 27.Jan.2000 (1)

Followup Exam 28.Apr.2003 (8)

a

b

Change:

Pooled std. dev.:

Change probability:

3 consecutives

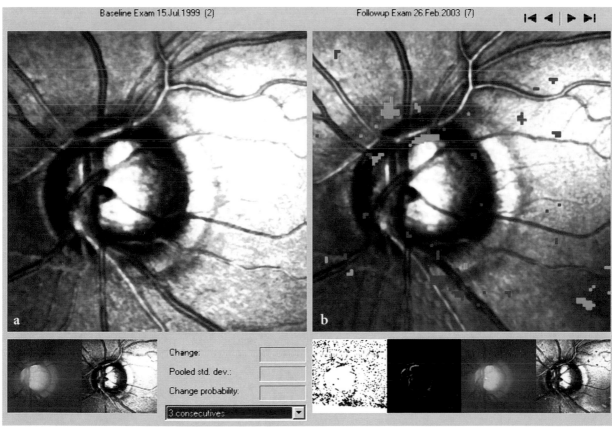

Baseline Exam 15.Jul.1999 (2)

Followup Exam 26.Feb.2003 (7)

a

b

Change:

Pooled std. dev.:

Change probability:

3 consecutives

5   Prominent Optic Discs

A. F. Scheuerle, E. Schmidt

5.1   Scanning of Prominent Discs

Laser scanning tomography was originally developed to evaluate and monitor glaucomatous optic disc cupping. However, prominent optic discs may also be examined using the HRT (Rohrschneider et al. 1990; Hudson et al. 1995; Göbel et al. 1997; Mulholland et al. 1998, Trick et al. 1998). The scanning process always runs from the anterior to the posterior parts of the eye. In the usual application, namely the diagnosis of glaucoma, the peripapillary retina is the most anterior structure to be exposed. For image acquisition of prominent discs, we suggest to focus not on the peripapillary retina but on the most prominent part of the disc instead. The camera of HRT II then performs an automatic pre-scan with 4–6-mm depth. From the images obtained in this pre-scan, the software computes and automatically sets the following parameters: the correct location of the focal plane, the required scan depth for that eye, and the proper sensitivity to obtain images with correct brightness.

The analysis of stereometric parameters always requires a reference plane. As the height of the temporal contour line segment might change in the course of papilledema the use of the standard reference plane is critical in follow-up exams. The mean retina height could serve as an alternative reference plane if the periphery of the field remains stable. We noticed that in papilledema the 15° field is often completely involved by swollen tissue. The original HRT requires manual adjustment of scan depth and sensitivity but allows for the selection of different fields of view. In our experience, the 20° field seems most suitable for quantitative and objective follow-up of papilledema (Scheuerle et al. 1999). It is very important to center the optic disc during image acquisition, otherwise, valuable parts of the field will be sacrificed for the alignment and normalization procedures. Depending on the underlying disease, short intervals between single examinations can be indicated.

Drusen of the optic disc may appear similar to papilledema in laser scanning tomographies. Analogous to clinical characteristics, we suggest looking for nodules and sharp disc borders as opposed to the smooth edges in papilledema. Using the macular edema module (MEM), we discovered that papilledemas usually demonstrate a wider parapapillary rim of increased signal width than do drusen of the optic disc.

Comments

- Prominent disc without excavation (1)
- Contour line (2) defines area of drusen which exceed the normal disc border
- Some drusen (3), which by no means posses a typical reflectivity signal, can be differentiated here based upon the relative increase in signal intensity resulting from their charaterstics shape at the surface of the disc
- Horizontal and vertical cross-sections demonstrate the prominence (4, 5) of this optic disc
- Variable form of the contour line height profile (6)

Case: A 63-year-old woman complained about decreasing visual acuity. Kinetic and static perimetry showed advanced visual field defects.

Original printout with:
a Topography image, OD
b Intensitiy image, OD
c Photograph of the optic disc, OD

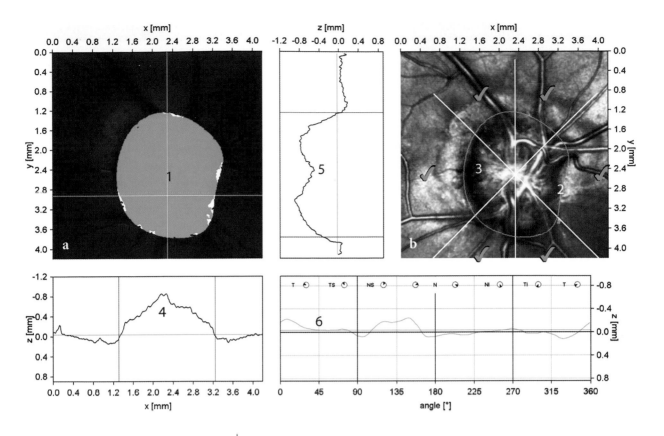

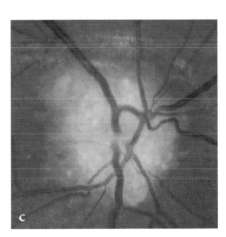

Stereometric Analysis ONH	
Disk Area	4.311 mm²
Cup Area	0.120 mm²
Rim Area	4.192 mm²
Cup Volume	0.005 cmm
Rim Volume	1.769 cmm
Cup/Disk Area Ratio	0.028
Linear Cup/Disk Ratio	0.166
Mean Cup Depth	0.009 mm
Maximum Cup Depth	0.028 mm
Cup Shape Measure	-0.246
Height Variation Contour	0.360 mm
Mean RNFL Thickness	-0.003 mm
RNFL Cross Sectional Area	-0.022 mm²
Reference Height	-0.036 mm
Topography Std Dev.	18 μm

Plate 63

Comments

- Photograph (a) and in particular intensity image (b) show slight papillary swelling (1) in the upper nasal quadrant
- Drusen are not identifiable in a conventional photo or in the intensity image
- Autoflorescence (c) clearly reveals papillary drusen (2) in the upper nasal quadrant
- The topography image (d) shows no excavation
- The three-dimensional reconstruction (e) confirms that the prominence is limited to the nasal parts of the disc

a Photograph of the optic disc, OD
b Intensitiy image, OD
c Autofluorescence, OD
d Topography image, OD
e 3D-image, OD

Plate 64

Comments

- The intensity image (b) shows a nasally pronounced papillary swelling (1)
- The prominence respects the borders of the optic disc
- The edema map (a) demonstrates only on the temporal disc border a rim of increased edema signal (2) which does not constitute intrapapillary fluid

a Edema map, OD
b Intensity image with interactive measurement, OD

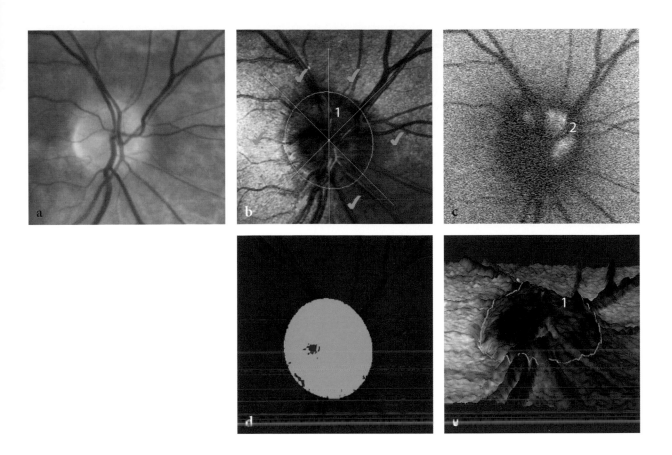

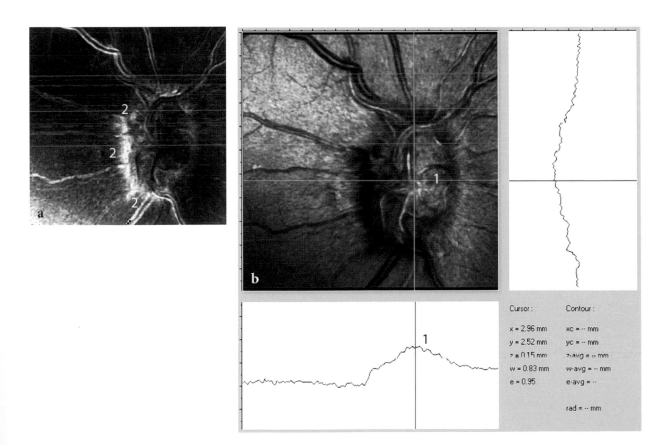

Plate 65

Comments

- Prominant optic disc (1)
- Disc borders of papillary drusen can be clearly defined by HRT image analysis (2) as opposed to traditional photographs where drusen color may simulate a blurred disk border (3)
- Only conventional color photograph clearly shows drusen (4)
- Thin rim of increased edema signal (5) which does not constitute intrapapillary fluid
- Drusen on the disc border influence the contour line height profile (d)

a Intensity image, OS
b Photograph of the optic disc, OS
c Edema map, OS
d Contour line height profile
e 3D-image, OS

Plate 66

Comments

- The intensity image (b) shows deeply placed drusen with weak prominance (1)
- Papillary borders (2) still identifiable

a Edema map, OD
b Intensity image of interactive measurement, OD

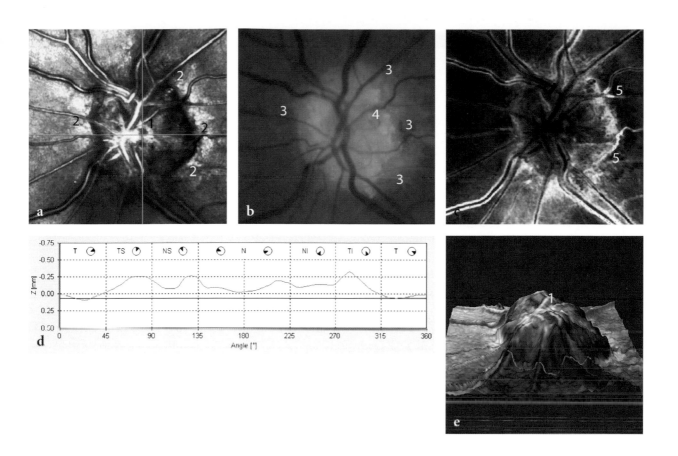

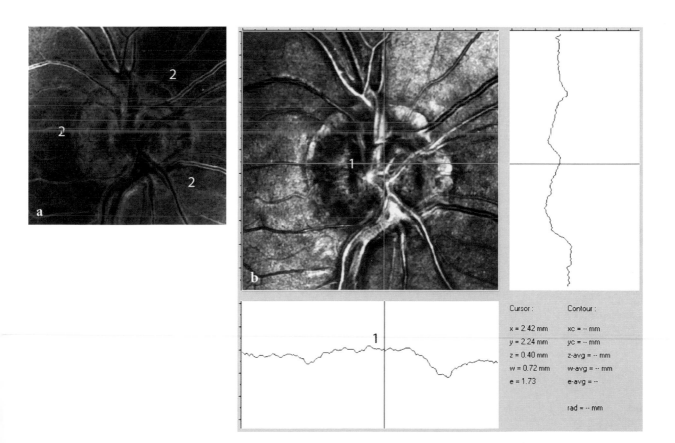

Comments

- Slightly swollen optic disc with decreased excavation (1)
- Decreased cup disk area ratio
- Rim volume artifically increased due to papillary swelling
- Falsely positive Moorfields classification (2) due to decreased excavation
- Papillary borders not well defined (3)
- Nasal prominence (4) in the horizontal cross-section
- No characteristic ("double hump") (5) form of the contour line (6)

Case: A 46-year-old female complained about blurred vision. Funduscopy showed bilateral swollen discs. Further examination revealed borreliosis. After appropriate antibiotic therapy, HRT measurements documented decreasing prominence of the optic discs.

Original printout with:
a Topography image, OD
b Intensitiy image, OD
c 30°-visual field, OD

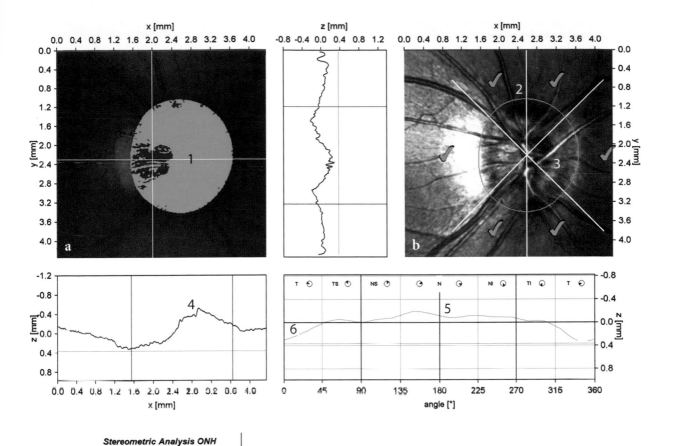

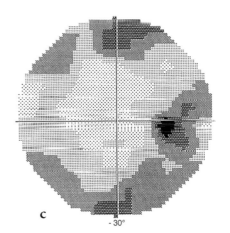

Stereometric Analysis ONH	
Disk Area	3.975 mm²
Cup Area	0.002 mm²
Rim Area	3.973 mm²
Cup Volume	0.000 cmm
Rim Volume	2.083 cmm
Cup/Disk Area Ratio	0.000
Linear Cup/Disk Ratio	0.020
Mean Cup Depth	0.031 mm
Maximum Cup Depth	0.107 mm
Cup Shape Measure	-0.205
Height Variation Contour	0.525 mm
Mean RNFL Thickness	0.371 mm
RNFL Cross Sectional Area	2.624 mm²
Reference Height	0.364 mm
Topography Std Dev.	20 μm

Plate 68

Comments

- The intensity image (b) reveals initial swelling of the optic disc (1) better than a traditional fundus phtograph (a)
- The intensity image of the retina shows a speckaled apprearance (3), possibly due to vitreal infiltration
- No visible excavation (2) in the horizontal cross-section

Case: A 37-year-old male with a history of Behcet's disease and recurrent posterior uveitis.

a Photograph of the optic disc, OS
b Intensity image with interactive measurement, OS

Plate 69

Comments

- Significant decrease of disc swelling (1) in subjective comparison of intensity images (a, b)
- Differential imaging analysis (c) objectively demonstrates the decrease (red color) in disc swelling (2)
- Corresponding follow-up photograph (e) shows an apparently normal optic disc with no signs of disc swelling
- Subjective comparison of horizontal cross-section images (d, f) demonstrates a significant decrease of papillary prominence
- However, a small amount of swelling is still detectable in the follow-up cross section image (f)

Note that the left eye (upper plate) of this patient presented only mild disc swelling that was documented using the HRT II 15° scan, while the right eye presented a more pronounced disc swelling and was therefore scanned with the HRT 20° field.

a 20° Baseline intensity image, OD
b 20° Follow-up intensity image, OD
c Topography difference image, OD
d Baseline horizontal crosssection
e Follow-up photograph, OD
f Follow-up horizontal cross section

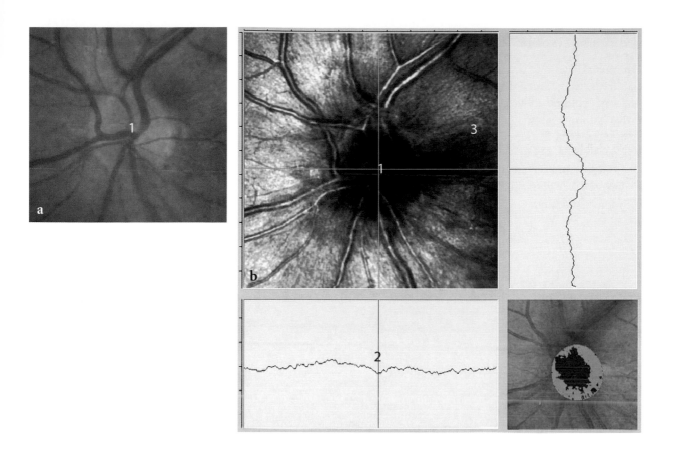

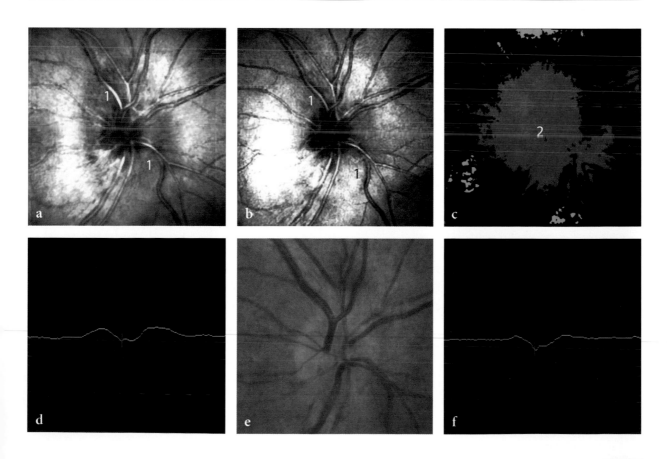

Plate 70

Comments

- Very large disc area value
- Large "volume above reference" and "volume above surface"
- No "volume below surface"
- Small central depression in the horizontal and vertical cross-section
- Contour line was used to mark the extension of papilledema and lies far away from the anatomic disc borders
- Therefore, contour line height profile can not be interpreted in the usual way

Case: A 23-year-old male presented with transient obscurations of vision. Magnetic resonance tomography revealed aqueductal stenosis and hydrocephalus.

Original HRT printout with:

a Topography image, OS
b Intensitiy image, OS
c 30°-visual field, OS
d Photograph of the optic disc, OS

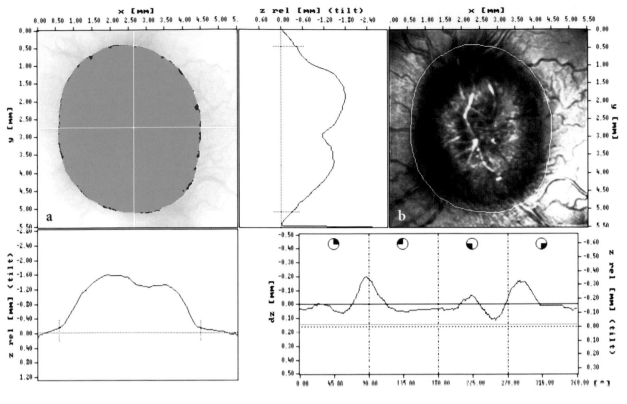

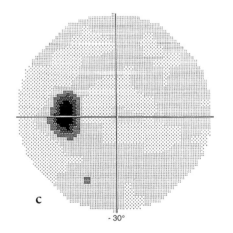

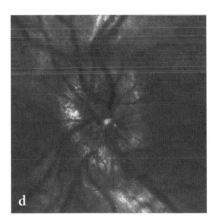

Stereometric Analysis :		Stereometric Analysis :	
Area:	15.018 mm²	Volume bel. Surface:	0.001 cmm
Effective Area:	0.172 mm²	Volume abv. Surface:	11.746 cmm
Area bel. Reference:	0.000 mm²	Mean Depth in Contour:	0.000 mm
Mean Radius:	2.188 mm	Effective Mean Depth:	0.008 mm
Mean Height of Contour:	-0.168 mm	Max. Depth in Contour:	0.031 mm
Height Variation Contour:	0.331 mm	Third Moment:	-0.438
Volume bel. Reference:	0.000 cmm	Reference Height:	-0.017 mm
Volume abv. Reference:	14.028 cmm		

Comments

- The topography difference image (a) demonstrates significant reduction of disc prominence
- Horizontal and vertical cross-sections confirm the decreasing disc prominence
- Subjective comparison of repeated follow-up intensity images (c, d, e) show a significant decrease of papilledema over a period of six months

Case: The young patient with aqueductal stenosis (previous plate) was initially treated with acetazolamid and received a ventriculoperitoneal shunt two weeks later. Fortunately, visual acuity of the left eye remained stable (20/30).

Original HRT printout with:
a Topography difference image, OS
b First follow-up (2 weeks) intensity image, OS
c Second follow-up image (1 month), OS
d Third follow-up image (3 month), OS
e Fourth follow-up image (6 month), OS

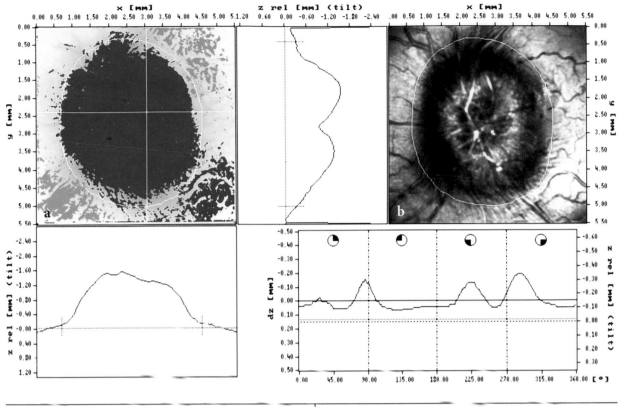

Stereometric Analysis :	Difference		Stereometric Analysis :	Difference	
Area:	15.018	+0.000 mm²	Volume bel. Surface:	0.002	+0.000 cmm
Effective Area:	0.270	+0.098 mm²	Volume abv. Surface:	10.067	-1.679 cmm
Area bel. Reference:	0.000	-0.000 mm²	Mean Depth in Contour:	0.000	0.000 mm
Mean Radius:	2.188	+0.000 mm	Effective Mean Depth:	0.007	-0.001 mm
Mean Height of Contour:	-0.157	+0.012 mm	Max. Depth in Contour:	0.029	-0.002 mm
Height Variation Contour:	0.261	-0.071 mm	Third Moment:	-0.494	-0.056
Volume bel. Reference:	0.000	-0.000 cmm	Referece Height:	-0.017 mm	
Volume abv. Reference:	12.184	-1.844 cmm			

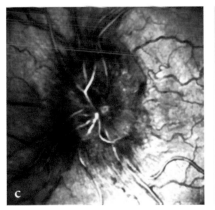

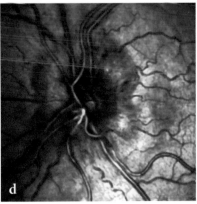

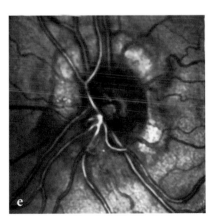

6 Macular Scans

A. F. Scheuerle, E. Schmidt

6.1 The Macular Edema Software Module

Since 1989, laser scanning tomography has been used to study the retinal surface in human eyes with macular pathologies (Bartsch et al. 1989). Laser scanning tomography provides precise measurements of the three-dimensional anatomy of macular holes and their rims for an objective evaluation of macular holes (Weinberger et al. 1995) and of the outcome of macular hole surgery (Hudson et al. 1997). In our experience, the postoperative closure of stage III or stage IV macular holes (as noted by clinical assessment) is usually indicated by significant changes in the topographic difference analysis, whereas, postoperative changes in stage II holes often proved to be too small to reach statistical significance.

Confocal laser scanning tomography is potentially useful as a noninvasive diagnostic technique for quantitative measurements of the neurosensory retinal detachment in central serous chorioretinopathy. Maximal heights in the geometric center of central serous chorioretinopathy blebs of up to 450 μm have been reported (Weinberger et al. 1996).

A few years ago it was discovered that not only retinal elevations but also retinal thickening could be measured using the HRT. The z-profile signal width analysis offers a non-invasive, objective, topographic, and reproducible index of macular retinal thickening (Hudson et al. 1998).

In the normal retina, the shape of the distribution of reflected light intensity along the optical axis, the confocal z-profile, is slightly asymmetric with a slightly longer tail towards deeper layers of the retina. This is due to light scattered from the deeper retinal tissue, which adds to the high reflectance of the internal limiting membrane. In the presence of a retinal edema, the amount of scatter inside the thickened retina highly increases, the z-profile becomes more asymmetric and its width increases (Fig. 14). At the same time, the peak reflectance at the internal limiting membrane is reduced, due to a decreasing change of refractive index at the retinal surface. Therefore, normalization of the z-profile width to the local reflectance produces an index for the likelihood of the presence of an edema:

$$e(x,y) = w(x,y) / r(x,y)$$

[$e(x,y)$ = edema index, $w(x,y)$ = z-profile width, $r(x,y)$ = normalized reflectance]

Figures 15–18a show the four types of images acquired by the Macular Edema Module (MEM). Figure 18b demonstrates the same image using the new Retina Module.

The new Retina Module now automatically excludes from further analysis all locations in the images that are over- or under-exposed or are invalid for other reasons. In addition, the average edema index values within nine zones centered on the fovea are computed. The nine zones comprise an inner circle of 0.5 mm radius and two rings with 1.0 and 1.5 mm outer radius. The two rings are each subdivided into four quadrants. The average edema index values in the nine zones can be displayed in a graph along the time axis for follow-up of the development of an edema.

The first studies were performed on patients with diabetic macular edema. In our experience, MEM and Retina Module are excellent tools for the diagnosis and follow-up of central serous chorioretinopathy and Irvine-Gass-syndrome. Examination is non-invasive and usually quick; however, image quality will be negatively affected by reduced transparency of the viewing optics. Until now, no validated correlation of retinal thickness measured by other techniques such as optical coherence tomography has been available for this new software. Interestingly, we have found different locations and shapes of local signal width and local reflectance measured by HRT compared with the leakage patterns of fluorescein angiography. This phenomenon might be explained by the fact that angiography is a dynamic process that locates extravasating dye while the MEM/Retina Module detect retinal thickening. Comparing both techniques, the examiner gets better information, not only where the fluid comes from, but also where intraretinal fluid accumulates in the long term.

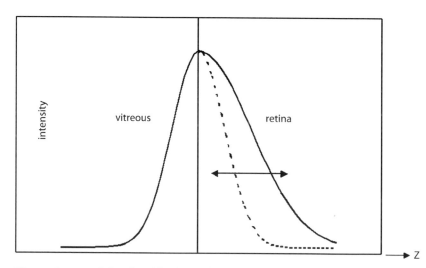

Fig. 14. Increased signal width of z-profile in edematous retina

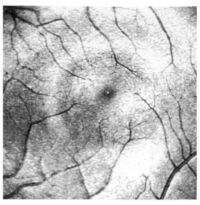

Fig. 15. Intensity image centered on a healthy macula

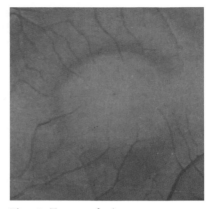

Fig. 16. Topography image

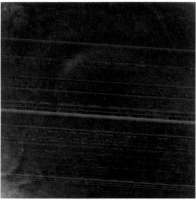

Fig. 17. Signal width image

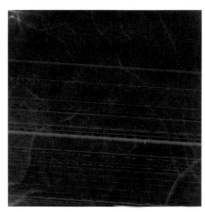

Fig. 18a. Edema map (MEM)

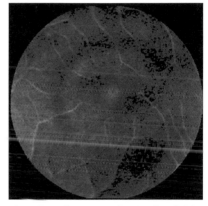

Fig. 18b. Edema map (Retina Module) in scale 0.0 – 4.0

Plate 72

Comments

- The intensity image (a) shows decreased retinal reflectivity. Here the macular area is marked by a circle (1)
- The signal width image (b) reveals radially arranged cystoid spaces (2) within the circled area
- The edema map (c) demonstrates the extension of macular edema
- Fluorescein angiography (d), 10 seconds after injection, showing near normal paramacular vessel anatomy
- Two minutes later (e) fluorescein angiography shows leakage (3) within the macular area
- Angiographic late phase is characteristic and reveals cystoid macular edema (4)
- Fluorescein angiography and HRT edema map display a similar extension of this macular edema

Case: A 54-year-old male presented with notable visual impairment three weeks after successful cataract surgery.

a Intensity image
b Signal width map
c Edema map in scale 0.0 – 4.0
d Fluorescein angiography (10 sec)
e Fluorescein angiography (2 min)
f Fluorescein angiography (12 min)

Plate 73

Comments

- The capillary phase of this fluorescein angiography (a) 10 seconds after injection clearly shows a regular pattern around the central avascular zone, the fovea still appears dark
- The late phase shows characteristic severe cystoid macular edema (b)
- The interactive mode with edema map demonstrates a very intense macular edema with clearly visible cystoid spaces in the center (c)
- Fluorescein angiography and HRT edema map display a similar extension of this macular edema
- Note that the irregular profile (1) does not reflect the retinal surface but expresses the amount of edematous retinal thickening
- Here, larger gaps (2) represent cystoid spaces surrounded from retinal tissue that is characterized by a large signal width and low reflectivity. The resulting e-value is therefore high represented by bright color in the edema map

Case: A 62-year-old female with rapid impairment of visual acuity two weeks after successful cataract surgery.

a Fluorescein angiography (10 sec)
b Fluorescein angiography (12 min)
c Edema map of interactive measurement

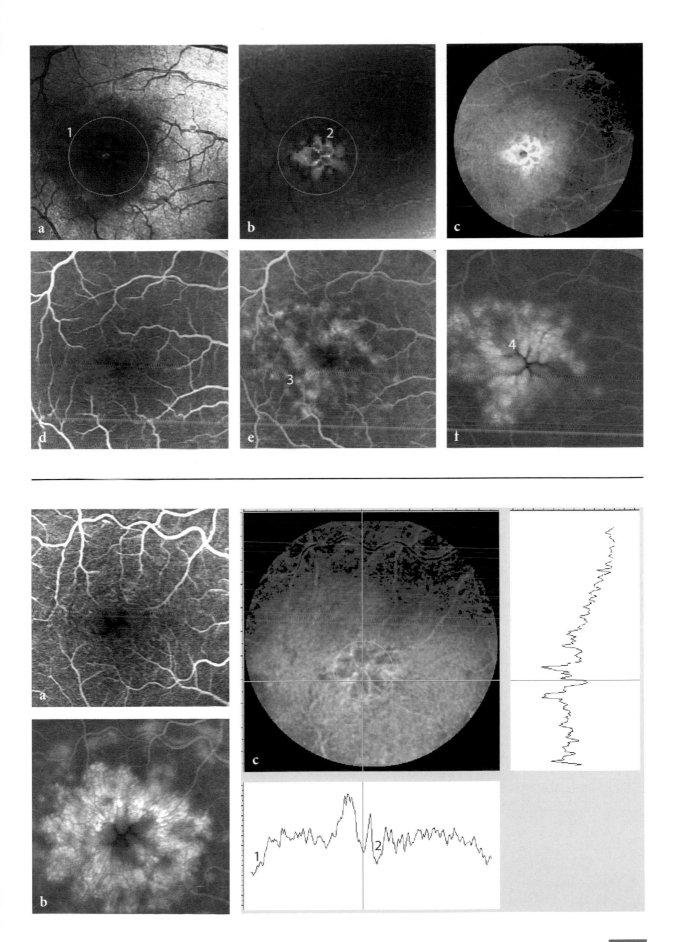

Plate 74

Comments

- Series of edema maps demonstrating the course of postoperative macular edema over time (d)
- Four weeks after uncomplicated cataract surgery visual acuity decreased to 20/60
- The macular edema rapidly decreased under anti-inflammatory treatment (b)
- However, a few days after dose reduction visual deterioration and recurrence of the macular edema were observed (c)
- Finally a stabe result was archieved (visual acuity 20/20)

Case: A 48-year-old patient with Irvine Gass syndrome. Eye exam and medical history revealed unilateral phacoplanesis due to blunt trauma twenty years ago.

Edema maps in scale 0.0 – 4.0:
a Baseline
b Follow-up after 36 days
c Follow-up after 57 days
d Long term follow-up graph
e Follow-up after 92 days

Plate 75

Comments

- The intensity image shows decreased retinal reflectivity of the macular region (1)
- Fluorescein angiography and HRT edema map display a similar extension of this macular edema regarding the maximum of intraretinal fluid collection (2)
- Only angiography demonstrates the foveal avascular zone (3) that is not identifiable in the HRT edema map
- Angiography reveals single points of leakage (4) whereas the HRT edema map pronounces areas of increased singal width i.e. intraretinal fluid
- The sector analysis (d) shows remarkebly higher e-values in the two nasal sectors of the edema map

a Intensity image
b Edema map in scale 0.0 – 4.0 with sector analysis
c Fluorescein angiography (15°)
d Table of sector analysis
e Fluorescein angiography (30°)

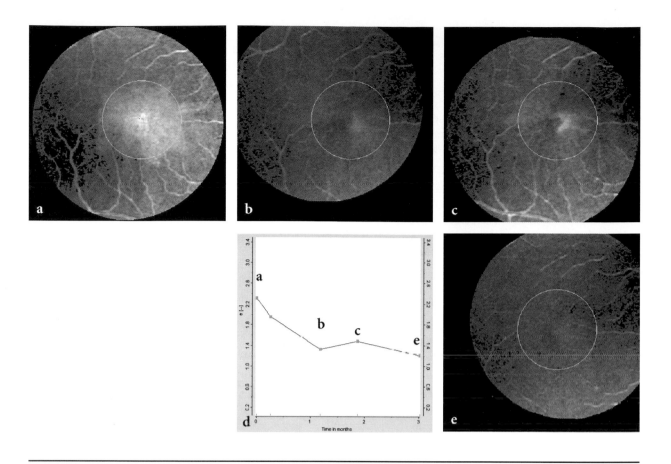

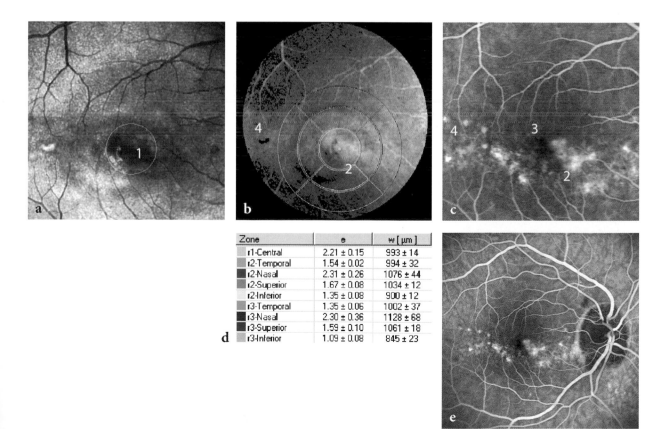

Zone	e	w [µm]
r1-Central	2.21 ± 0.15	993 ± 14
r2-Temporal	1.54 ± 0.02	994 ± 32
r2-Nasal	2.31 ± 0.26	1076 ± 44
r2-Superior	1.67 ± 0.08	1034 ± 12
r2-Inferior	1.35 ± 0.08	900 ± 12
r3-Temporal	1.35 ± 0.06	1002 ± 37
r3-Nasal	2.30 ± 0.36	1128 ± 68
r3-Superior	1.59 ± 0.10	1061 ± 18
r3-Inferior	1.09 ± 0.08	845 ± 23

Comments

- Fluroescein angiography demonstrates a large hyperfluorescent zone in the perifoveal area (1)
- The intensity image shows decreased reflectivity (2)
- The HRT edema map reveals edema maximum just below the foveola (3)
- The follow-up intensity image demonstrates increased reflectivity (4) and reduced edema (5) eight weeks after laser treatment. The follow-up edema map now displays a more diffuse retinal edema
- Formation of a stage III macular hole could be suspected in the HRT interactive measurement mode (8) but clinical examination and optical cohererence tomography demontrate a layered macular pseudohole (6) resulting from long-lasting cystoid macular edema
- However, the thin layer is not detected by the automatic surface line of the OCT (7)

Case: 62-year-old male presented with venous branch occlusion, severe macular edema and reduced visual acuity. He was treated with focal argon laser coagulation. Despite decreased macular edema, his visual acuity did not improve.

a Fluorescein angiography (12 min)
b Baseline intensity image
c Baseline edema map in scale 0.0 – 4.0
d Follup-up intensity image
e Follow-up edema map in scale 0.0 – 4.0
f Optical coherence tomography (OCT)
g Automatic surface line of the OCT
h Topography image of interactive measurement

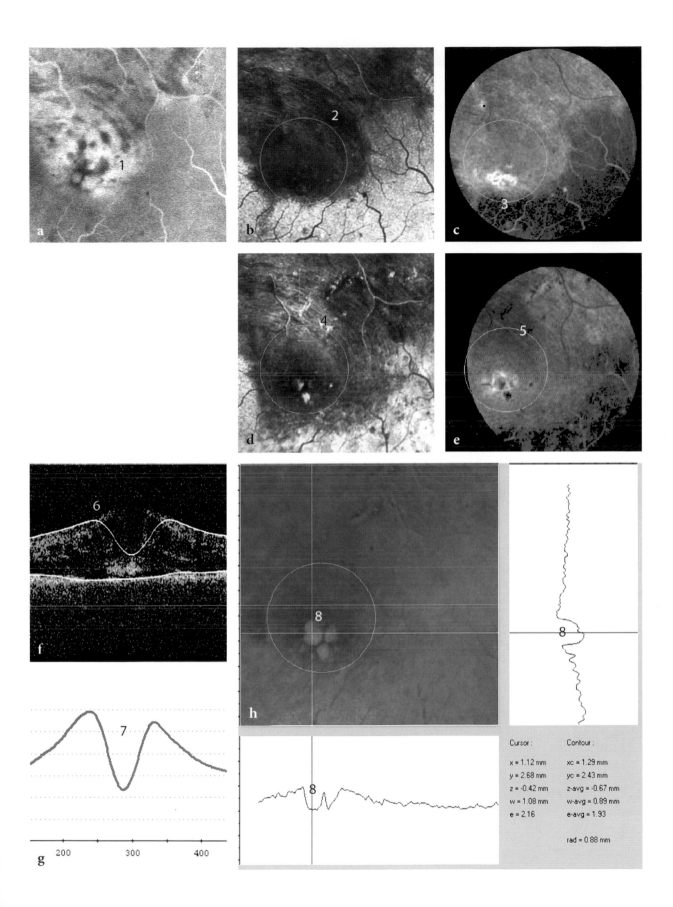

Comments

- The intensity image shows a markedly decreased reflectivity in the macular area (1)
- The edema map presents its maximum intensity (2) just below the foveola
- Fluorescein angiography with an isolated point of leakage (3) just above the foveal avascular zone
- Follow-up measurements were performed every three weeks and initially document an enlargement of macular edema (d, e) with subsequent reduction overtime (g, h, j, k)
- Optical coherence tomography initially demonstrates subretinal fluid and detachment of the retinal pigment epithelium (i). The follow-up examination showed a slight foveal depression indicating less subretinal fluid and confirming the results of HRT edema maps

Case: A 36-year-old male with ideopathic central serous retinopathy. Micropsia and visual acuity slowly improvement over two months.

a Baseline intensity image
b Baseline edema map in scale 0.0 – 4.0
c Baseline fluorescein angiography
d First follow-up intensity image
e First follow-up edema map in scale 0.0 – 0.4
f First follow-up fundus photograph
g Second follow-up intensity image
h Second follow-up edema map in scale 0.0 – 0.4
i Second follow-up optical coherence tomography (OCT) of the macula
j Third follow-up intensity image
k Third follow-up edema map in scale 0.0 – 0.4
l Third follow up OCT

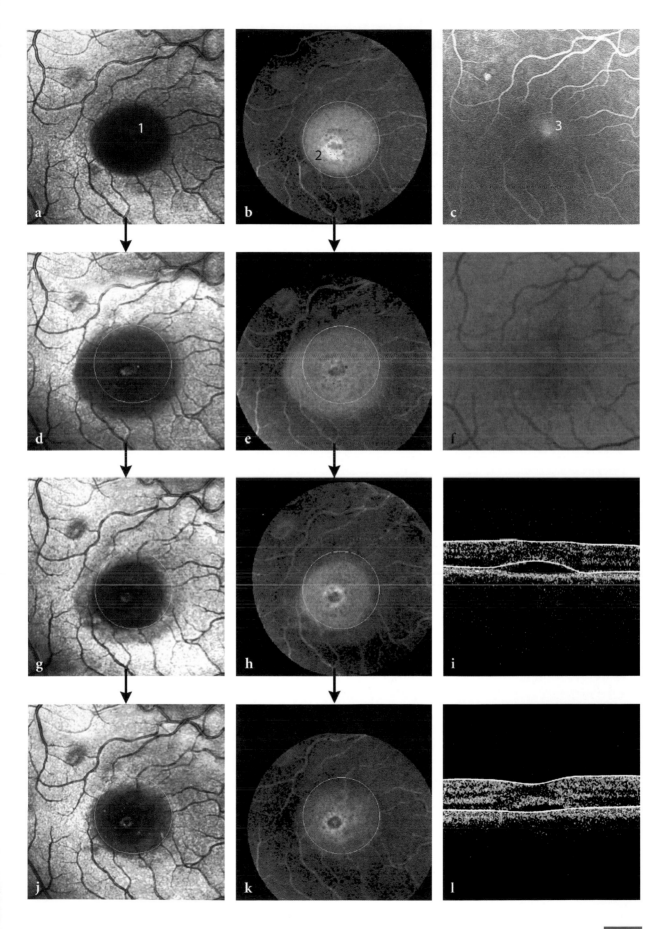

Plate 78

Comments

- Fluorescein angiography reveals several spots of choroidal hyperpermeability (1) whereas the HRT edema map shows a large area of diffusely increased signal width i.e. intraretinal fluid affecting the temporal central region (2)
- Intensity image shows decreased retinal reflectivity of the macular region (3)
- The HRT edema map indicates the greatest level of edema in the foveal region (4), this is not visible during the mid-phase of angiography
- The follow-up intensity image demonstrates significantly increased reflectivity (d) after three weeks
- HRT edema map appears almost normal (e) after three weeks, correlating with an improvement in visual acuity
- Black edges (5) are artifacts secondary to alignment of follow-up images

Case: A 35-year-old male with central serous retinopathy and reduced visual acuity which significantly improved after three weeks.

a Fluorescein angiography (15°)
b Baseline intensity image
c Baseline edema map
d Follup-up intensity image
e Follow-up edema map

Plate 79

Comments

- The intensity image shows a dome-shaped area of decreased retinal reflectivity (a)
- The edema map presents its maximum signal intensity directly in the macula (b)
- Fluorescein angiography with a comparable area of diffuse leakage (c)
- The most intense leakage can be observed in the macula (1)
- Fundusphotography (d) shows the same shape of detachment of the retinal pigment epithelium as in HRT images, angiography and autofluoresecence (f). The yellow granulas are best visible in color photograph
- Also the signal width (e) map demonstrates its maximum signal intensity in the foveal region

Case: A 42-year-old male with chronic central serous retinopathy.

a Intensity image (15°)
b Edema map (15°) in scale 0.0 – 4.0
c Fluorescein angiography (20°)
d Fundusphotograph (20°)
e Signal width map (15°)
f Autofluorescence (20°)

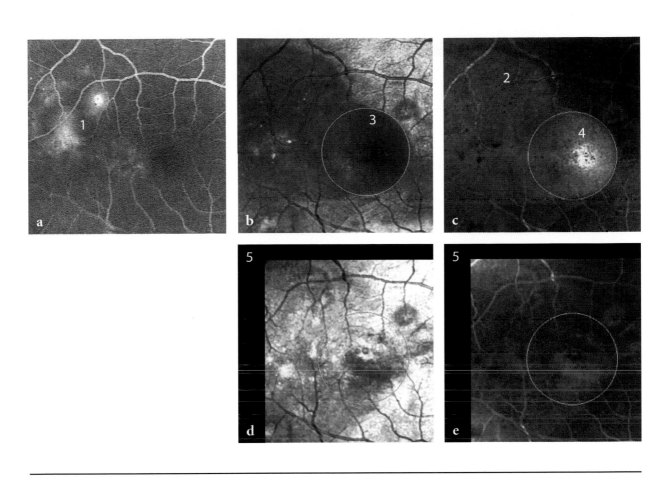

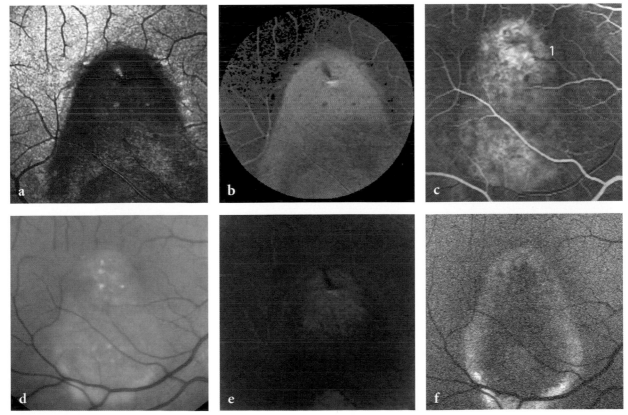

Plate 80

Comments

- The intensity image shows a perifoveal pattern (1) of decreased reflectivity (a)
- The edema map and signal width map confirm this pattern (b, c)
- Optical coherence tomography (d) and HRT topography image (f) reveal increased retinal thickness and prominence
- Fluorescein angiography shows only streched perimacular vessels

Case: A 55-year-old male with metamorphopsia and decreasing visual acuity secondary to a macular pucker.

a Intensity image
b Edema map
c Signal width map
d Optical coherence tomography
e Fluorescein angiography (1 min)
f Topography image

Plate 81

Comments

- The intensity image (a) demonstrates retinal folds of decreased reflectivity running in the direction of the macula
- The edema map (b) enhances the visibility of these folds
- The HRT edema map (b) and fluorescein angiography (e) display a similar extension of macular edema (1, 2)
- Fundusphotograph (d) displays a more pronounced cellophanelike aspect (3) of the retinal surface
- Fluorescein angiography shows several points of leakage wheras retinal folds are more difficult to see

Case: A 44-year-old female with bird shot syndrome and severe epiretinal gliosis.

a Intensity image
b Edema map
c Signal width map
d Photograph of the macula
e Fluorescein angiography of the macula
f Topography image

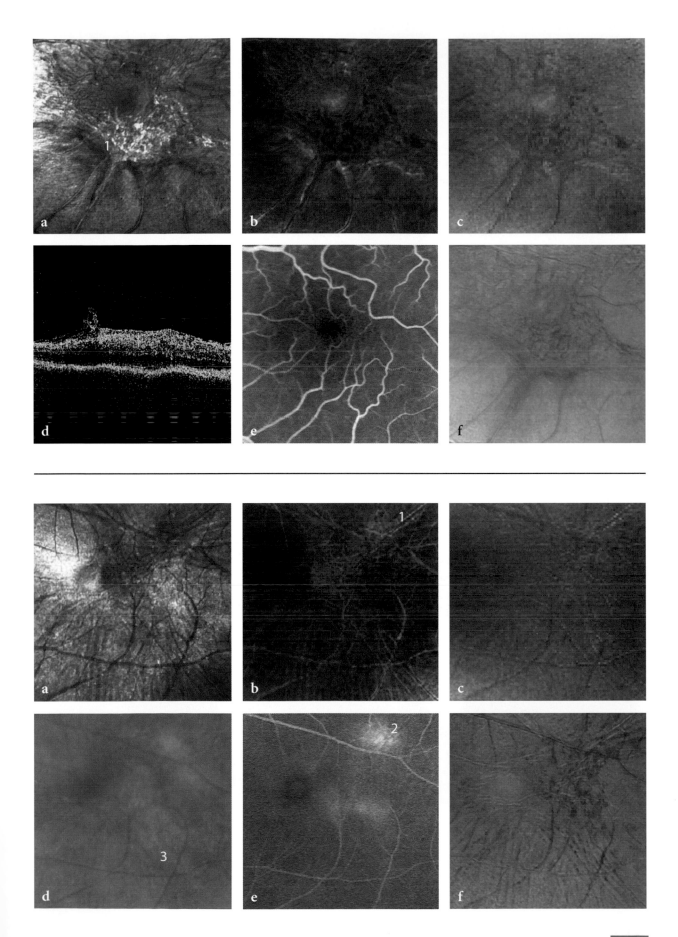

Plate 82

Comments

- The topography image shows a large circular macular hole (a)
- The area and volume below the reference plane are interpreted as "cup"-parameters (1) by the HRT software
- The intensity image presents decreased reflectivity of the perimacular area (2) with increased intensity at the macular hole
- The base of the macular hole is completely flat (3)
- The height profile of the contour line displays a symmetrical rim
- Autofluorescence using blue laser light (c) demonstrates horizontal and vertical dimensions similar to HRT topography but can not measure depth

Case: A 62-year-old female with severly reduced visual acuity secondary to a macular hole.

Original HRT I printout with:
a Topography image
b Intensity image
c Autofluorescence (15°)

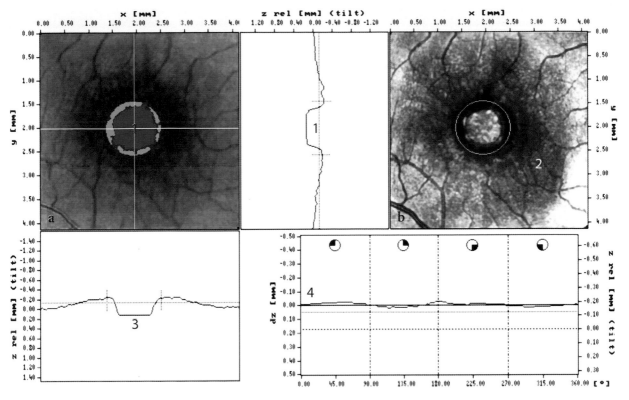

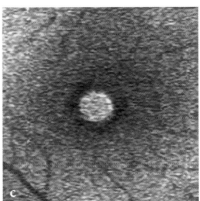

Stereometric Analysis ONH:

Disk Area:	1.017	mm²
Cup Area:	0.643	mm²
Cup/Disk Area Ratio:	0.632	
Rim Area:	0.375	mm²
Cup Volume:	0.138	cmm
Rim Volume:	0.021	cmm
Mean Cup Depth:	0.221	mm
Maximum Cup Depth:	0.339	mm
Cup Shape Measure:	0.255	
Height Variation Contour:	0.054	mm
Mean RNFL-Thickness:	0.052	mm
RNFL-Cross Section Area:	0.185	mm²
Reference Height (Std.):	-0.127	mm

Plate 83

Comments

- The intensity image (c) shows decreased reflectivity of the perimacular area (1) while the macular hole itself appears bright
- The topography image (a) demonstrates a (very) bright (deep) cup area. The macular hole has slightly less intense signal surrounded by a dark (prominent) ring (2)

Case: A 68-year-old female with severly decreased visual acuity seondary to a macular hole.

a   Topography image (20°)
b   Horizontal crosssection
c   Intensity image (20°)

Plate 84

Comments

- The edema map (a) demonstrates retinal folds with increased signal intensity indicating epiretinal gliosis
- A ring-shaped structure of high signal intensity (1) can be observed in the foveal region
- The topography image shows a bright (deep) structure (2) in the foveal region
- Interactive measurement (c) identifies a macular hole (3)

Case: A 54-year-old female presented with significantly decreasing visual acuity and metamorphopsia.

a   Edema map
b   Topography image
c   Intensity image of interactive measurement

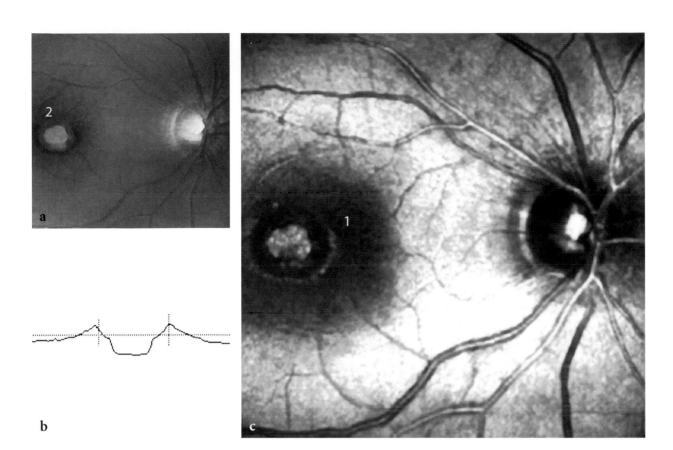

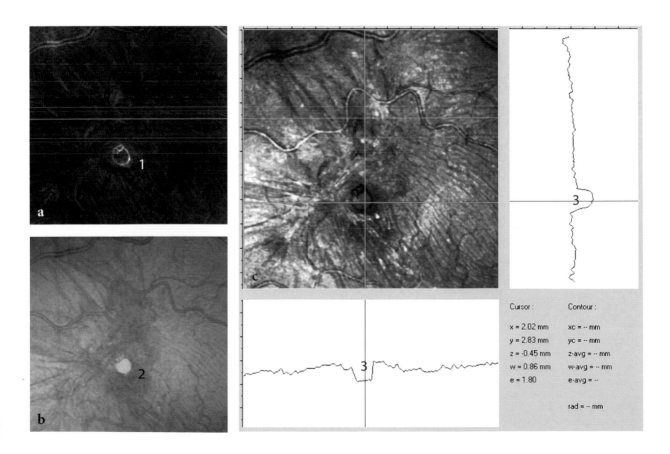

Comments

- The intensity image (a) shows decreased reflectivity of the perimacular area (1) while the macular hole itself appears bright (2)
- Horizontal and vertical cross-sections demonstrate the elevated perifoveal area (3)
- A small floater of the vitreus body (4) leads to a slight depression of the surface profile on the nasal edge of the macular hole
- The topography image (b) demonstrates a very bright (deep) macular hole area (5) surrounded by a dark (prominent) ring
- The edema map (d) reveals some more details of the perifoveal structures. The small floater of the vitreus body appears very intense (6) and simulates an extraordinary amount of intraretinal fluid (7)
- Optical coherence tomography (OCT) (e, f) demonstrates a macular hole with symmetrically detached perifoveal retina similar to HRT cross-sections

Case: A 59-year-old female with decreasing visual acuity seondary to a macular hole.

a Intensity image of interactive measurement
b Topography image
c Fundusphotograph
d Edema map of interactive measurement in scale 0.0 – 4.0
e Optical coherence tomography (OCT) of the macula
f Automatic surface line of the OCT

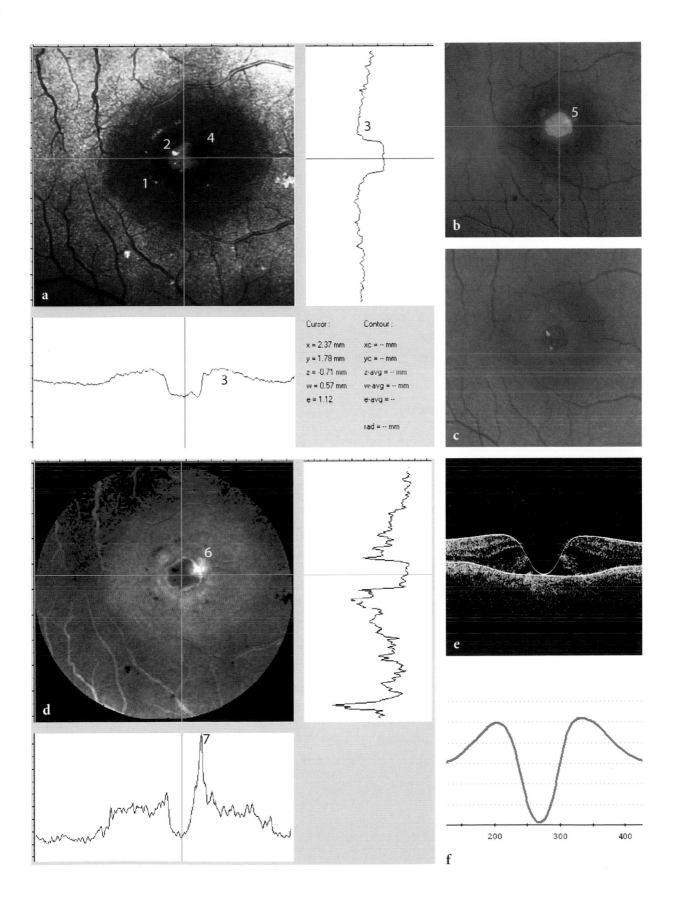

Cursor:

x = 2.37 mm
y = 1.78 mm
z = -0.71 mm
w = 0.57 mm
e = 1.12

Contour:

xc = ·· mm
yc = ·· mm
z-avg = ·· mm
w-avg = ·· mm
e-avg = ··

rad = ·· mm

Plate 86

Comments

- The intensity image (a) shows decreased reflectivity of the perimacular area (1) while the macular hole itself (2) appears bright
- The topography image (b) demonstrates a bright (deep) macular hole surrounded by a dark (prominent) ring with cystoid formations (3)
- The enlarged horizontal cross-section (c) shows the macular hole with perifoveal elevation
- The depression in the heigth profile next to the macular hole represents one of the surrounding cysts (4)

Case: A 67-year-old male presented with decreasing visual acuity seondary to a macular hole.

a Intensity image
b Topography image
c Horizontal crosssection

Plate 87

Comments

- The baseline intensity image (a) shows decreased reflectivity of the perimacular area (1) while the macular hole itself (2) appears bright
- The baseline topography image (d) demonstrates a bright (deep) macular hole surrounded by a dark (prominent) ring with cystoid formations (3)
- The first follow-up of intensity and topography images reveals a significant enlargement of the macular hole and the surrounding retinal elevation (b, e) while the cysts are not visible any longer
- Both second follow-up images (c, f) show a closed macular hole (4, 5) one year after successful vitreoretinal surgery

Case: A 70-year-old male was diagnosed with a macular hole but refused a surgical intervention for six months because of initially satisfying visual acuity.

a Baseline intensity image
b First follow-up intensity image
c Second follow-up intensity image
d Baseline topography image
e First follow-up topography image
f Second follow-up topography image

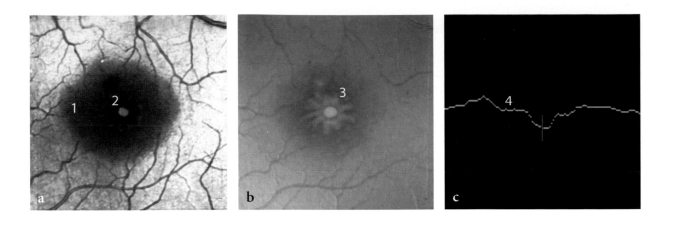

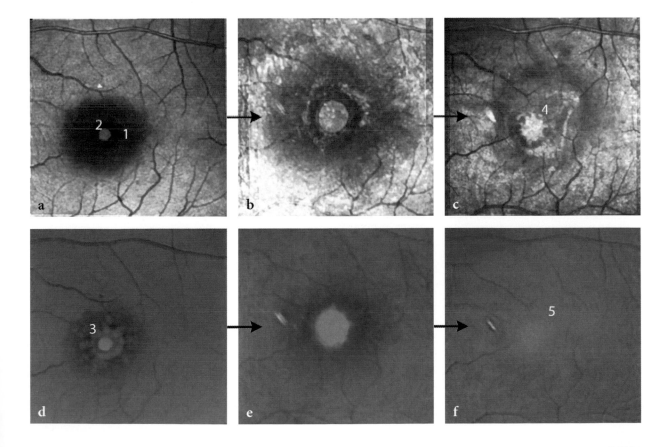

Plate 88

Comments

- The baseline intensity image (a) shows a large macular hole with a surrounding area of reduced reflectivity
- The baseline topography image (c) demonstrates a deep (bright) macular hole and perifoveal elevation (dark ring)
- Autofluorescence (e) using blue laser light demonstrates a circular hyperreflective area corresponding to bare retinal pigment epithelium which appears slighly smaller than the macular hole in HRT images
- The early phase of fluorescein angiography (g) shows some points of leakage in the macular area, but also the late phase (not shown here) does not reveal characteristics of cystoid edema
- In this patient, neither HRT images nor angiography show any cystoid macular formation
- All follow-up images (b, d, f, h) show a closed macular hole six months after successful vitreoretinal surgery

Case: A 50-year-old female underwent successful vitreoretinal surgery for a macular hole and regained good visiual acuity.

a Baseline intensity image
b Follow-up intensity image
c Baseline topography image
d Follow-up topography image
e Baseline autofluorescence image (15°)
f Follow-up autofluorescence image (15°)
g Baseline fluorescein angiography (15°)
h Follow-up fluorescein angiography (15°)

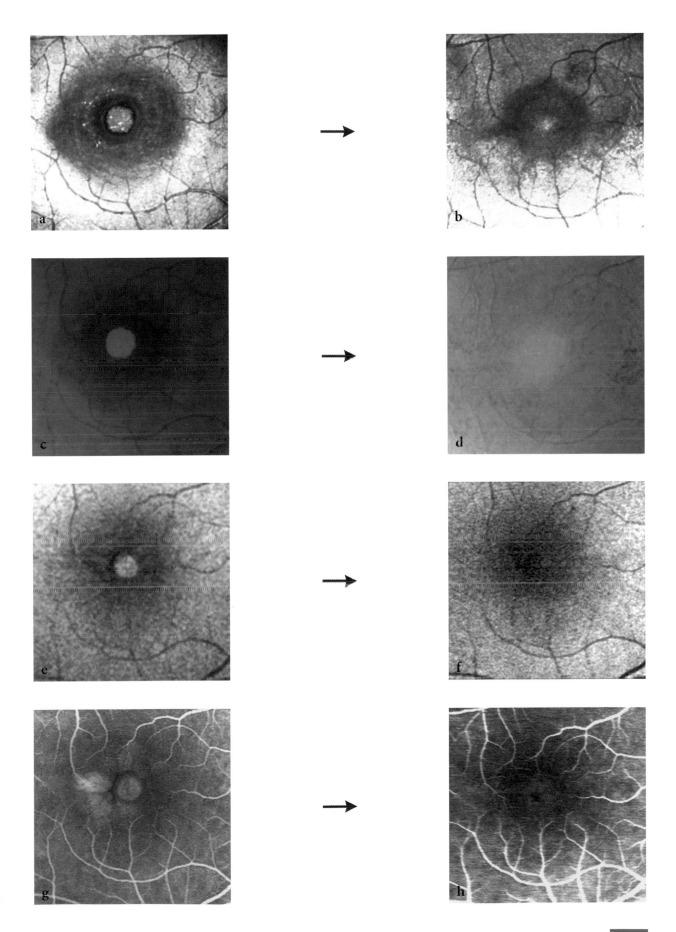

7 Miscellaneous

A. F. Scheuerle, E. Schmidt

7.1 Optimizing Imaging

The HRT user will occasionally encounter images that are difficult to interpret or are even misleading. Irregularities of the cornea caused by astigmatism or dry eye symptomatic are the most frequent reasons for poor image quality besides cataract. A high standard deviation (>30) indicates increased signal noise, producing inaccuracies in baseline and follow-up results. An intact tear film is a prerequisite for good image quality in laser scanning tomography. We found hyaluronic acid and carbomer more suitable than methylcellulose to improve image quality in patients with dry eye symptomatic. In any case, repeated blinking between the recording of image series is mandatory. Astigmatism might be compensated by the algorithm of the HRT software up to a certain extent. Sheen et al showed that optic disc parameters were not significantly affected by induced astigmatism of up to 2.5 dpt and concluded that astigmatism up to 2.5 dpt would not require correction (Sheen et al. 2001). Having evaluated many tomographies affected by astigmatism we can only partially confirm these results. This study pointed out that both the standard deviation of the mean topographic images and the z-profile half-maximum width of the axial intensity profile were significantly greater with induced astigmatism of ≥3.0 dpt. From our personal experience, astigmatism >2.0 dpt will lead to significantly decreased image quality. Therefore, we recommend to correct astigmatism >1.0 dpt. The patient's spectacle glasses might be used for correction, but progressive lenses complicate reproducibility and glasses often produce disturbing reflections. If there is a choice, we would prefer contact lenses instead, which may even correct some irregular astigmatism. Although the mechanism of magnetic adhesion has not been completely developed, astigmatic lenses that can be directly attached to the HRT II eyepiece represent the most adequate correction for the majority of patients. In any case of astigmatic correction, the examiner has to make a permanent note about the technique and amount of correction used to generate similar conditions for follow-up imaging. Orgül et al. described that misalignment errors between the patient and the laser scanner may account for significant variability with the HRT (Orgül et al. 1996). Therefore, a comfortable but also stable standard position of the patient's head has to be established.

The principle of equal optical conditions should be applied for imaging eyes that undergo cataract or refractive surgery, too. As intraocular magnification might change after such procedures, a new baseline instead of further follow-up exams is recommended. The HRT II software has enhanced capabilities to compensate different working distances compared with the original HRT. Problems with inadequate normalization and poorly fitting contour lines in follow-up images have almost been eliminated. Generally, a working distance of 15 mm is reasonable for the 15° field of HRT II. Shorter distances may irritate the patient, with the HRT front lens getting in contact with lashes and becoming dirty. Larger distances decrease the intensity of laser light passing through regular undilated pupils and provoke shades in the periphery of the tomography as shown in Fig. 19. Although detector sensitivity automatically increases, overall image quality certainly decreases under these conditions. Even more important, extreme differences of the working distances between the baseline and follow-up image acquisition might cause an incorrect placement of the contour line in the follow-up image due to incompatible magnification factors (Figs. 20–22).

Despite a centered beam and a correct working distance, laser light may be blocked by narrow pupils, especially under pilocarpine medication. In these cases, pharmaceutical dilation of the pupil is usually necessary to obtain acceptable image quality. Floaters of the vitreous body or the ring of Martegiani are other common intraocular conditions that can be responsible for underexposed areas within the tomography and may even cause misleading results.

Finally, we want to remind the HRT-user that changes in optic disc parameters can be correlated to changes in intraocular pressure (IOP) between follow-up exams (Lesk et al. 1999). Even moderate IOP differences (>8.0 mm Hg) might significantly affect disc topography and should be avoided if possible (Bowd et al. 2000, Kotecha et al. 2001).

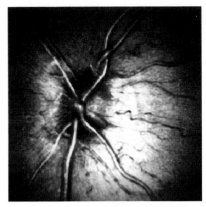

Fig. 19. HRT II intensity image with shaded periphery due to a greater than normal working distance

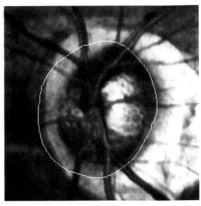

Fig. 20. HRT intensity image showing an incorrect nasal "enlargement" of the glaucomatous optic disc secondary to a greater than normal working distance during the follow-up image acquisition

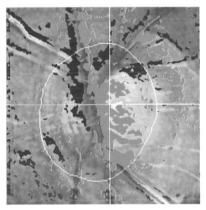

Fig. 21. HRT follow-up image with surface height difference analysis

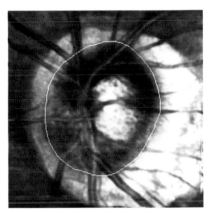

Fig. 22. HRT intensity image of the same eye with correct optic disc properties

Comments

- The intensity image and 3-dimensional reconstruction clearly demonstrate high unspecific signal that leads to decreased image quality (1)
- The topography image presents with an unusual orange color, vessels appear bright instead of dark (2)
- Proper astigmatic correction increases image quality significantly (3)
- Topography image can reliably detect cup parameters (4)
- A significant reflection artifact caused by patient's corrective glasses simulates a circumscribed retinal prominence (5)

Plate 89

a 3D-image, regular scan, OS
b Topography image, regular scan, OS
c Intensity image, regular scan, OS
d 3D-image, astigmatic correction, OS
e Topography image, astigmatic correction, OS
f Intensity image, astigmatic correction, OS

Comments

- The intensity image and 3-dimensional reconstruction demonstrate a circumscribed intrapapillary prominence (1)
- The topography image shows an unusual green "island" within the cup area (2)
- Patient's corrective glasses were effectivly used to perform astigmatic correction but caused a significant reflection artifact that simulated increased rim parameters and could affect the classification of this optic disc

Plate 90

a Topography image, OS
b Intensity image of interactive measurement, OS
c 3D-image, OS

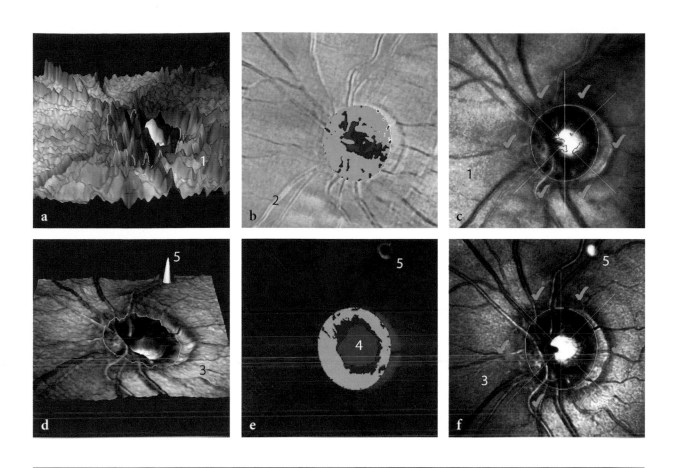

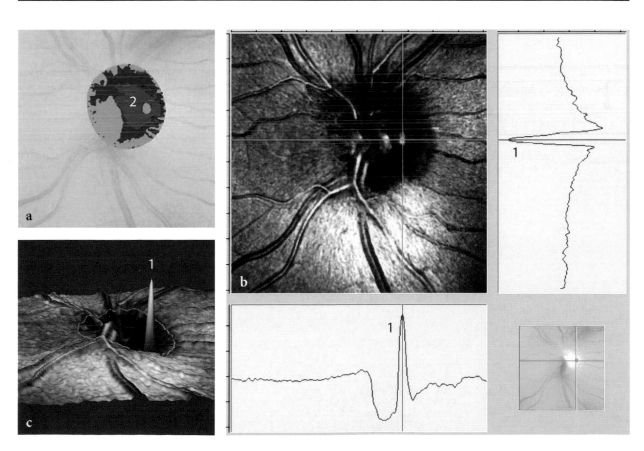

Plate 91

Comments

- Fundusphotograph (a) shows two subretinal PFCL bubbles after intraocular surgery for retinal detachment (1)
- The subretinal bubbles appear as dark rings in the HRT intensity image (2)
- Abnormal reflectivity may lead to an artificial prominence (or depression) in the retinal surface profile (3)
- Horizontal cross-section image shows a signal (3) similar to those generated by prominent vessels (4)

a Fundusphotograph, OD
b Intensity image of interactive measurement, OD

Plate 92

Comments

- HRT intensity image (b) demonstrates medullated nerve fibers as bright structures (1), clearly different from normal retina
- The reflectivity pattern of medullated nerve fibers leads to a small prominence of this area in the retinal surface profile (2), confirmed by 3-dimensional reconstruction (3)

a 3D-image, OD
b Intensity image of interactive measurement, OD

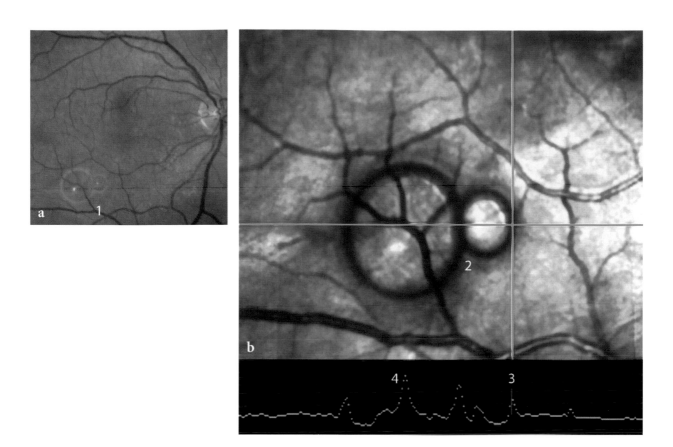

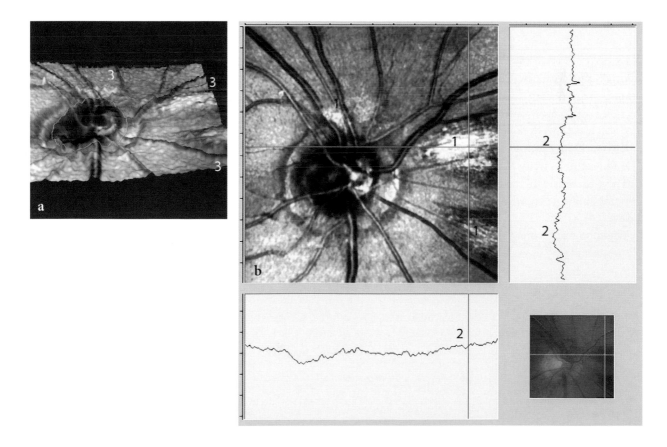

Plate 93

Comments

- Parapapillary shade caused by a vitreous floater (1)
- Abnormal reflectivity may lead to artificial depression (or prominence) in the retinal surface profile (2)
- Neither horizontal nor vertical cross-section images show a signal originating from the vitreus body (3)
- 3-Dimensional reconstruction demonstrates a parapapillary "pit" (4)

a   Topography image, OD
b   Intensitiy image of interactive measurement, OD
c   3D-image, OD

Plate 94

Comments

- Ring of Martegiani causes an unusual reflectivity pattern and appears dark (1) in the HRT intensity image (b)
- Abnormal reflectivity may lead to artificial prominence (or depression) in the retinal surface profile (2)
- Vitreus floater presenting with similar color (3) but a significantly different reflectivity pattern
- The 3-dimensional reconstruction (c) demonstrates a prominent parapapillary ring while the vitreus floater appears to be at the retinal level (4)
- The topography image (d) is hardly affected by the extrapapillary ring of Martegiani

a   Photograph of the optic disc, OD
b   Intensity image of interactive measurement, OD
c   3D-image, OD
d   Topography image, OD

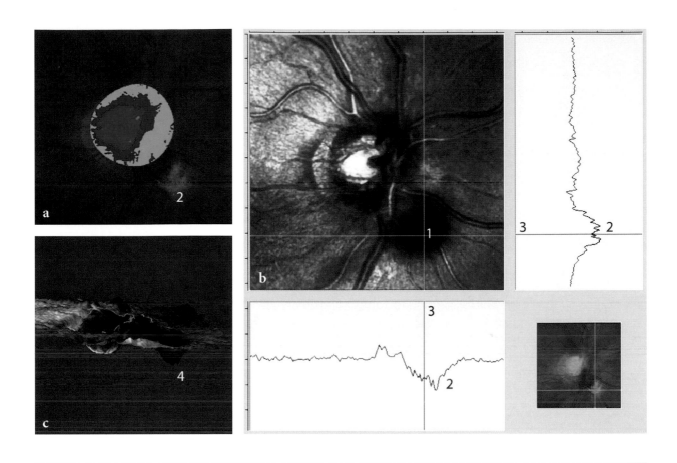

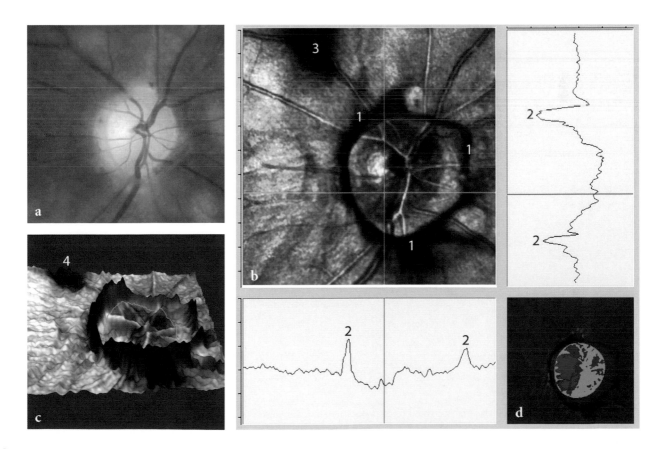

8 Confocal Corneal Imaging

J. Stave, R. Guthoff*

* Universitäts-Augenklinik, Doberaner Str. 140, 18057 Rostock, Germany

8.1 The Rostock Cornea Module

Analog-digital, confocal scanning slit microscopes are used for in vivo examinations of the cornea, e.g., following LASIK to obtain images providing morphological data regarding wound repair in the epithelium, reinnervation and the behavior of the keratocytes in the keratome incision area over an extended period. Due to the loose optical coupling to the cornea by means of a gel there are limitations on the information that can be obtained regarding the exact depth location of the optical section in the cornea, which is needed for a three-dimensional reconstruction of the cell structures. The electromechanical scanning slit method entails also systematic errors, such as non-uniform image illumination and image distortion.

The HRT II was modified by adding a detachable objective system, the so-called "Rostock Cornea Module", combined with a fine-thread or hydraulic z-drive to shift the focal plane in the cornea as a starting plane for the optionally available internal z-scan. During the examination, the distance between the cornea and the microscope is kept stable by a contact element that is optically coupled to the eye only by the lacrimal film or a protective gel. This contact element consists of a thin acrylic disc located between the objective and the cornea. Depending on the objective combination selected and the maximum internal z-scan (4 mm), it is possible to image a depth field of up to 30 µm. By deactivating the internal z-scan and manually adjusting the initial z-setting on the Rostock Cornea Module to a desired depth – the LASIK incision area for example – a series of images can be collected from this depth with an image yield of approximately 100 % and an exact depth allocation. The HRT II automatic threefold series imaging and counting software can be used to average up to 96 images, for instance, to determine the keratocyte density in the stroma. An HRT II novelty is that the z-shift between images is stepped, i.e., the z-value remains constant during imaging. This is a major innovation and a prerequisite for producing images of structures in a single plane without any distortion in direction z. Fulfilling this crucial prerequisite for producing distortion-free, three-dimensional reconstructions is a landmark achievement.

A long focal length, water-immersion objective with a large aperture was selected to obtain a high magnification. In addition, corrected lenses are attached to the HRT II objective to reduce the laser scanning angle to approximately 7.5° (fixed at 15° on the HRT II) to attain the necessary magnification, i.e., the scanning field is reduced, increasing the magnification. The size of the scanning field in the object is then 250–250 μm. Additional lenses allow an area of up to 500–500 μm to be scanned. Video monitoring further facilitates the exact vertical positioning of the cornea in front of the microscope within the micrometer range. Microscopy of the cornea at up to 800fold magnification is possible and effective.

The Rostock Cornea Module on the HRT II allows high-contrast imaging of the layered structure of the ocular epithelium. Due to the good depth resolution of the microscope, optical sections that are only a few micrometers thick can be imaged and precisely measured. This also applies to the nerve plexus, the entire stroma including the keratocytes and the endothelium with its microstructure. Plate 95c–h shows confocal laser scanning micrographs of normal and phathological structures of the cornea.

The Rostock Cornea Module/HRT II combination enables unproblematic in vivo examination of the anterior section of the eye, including the cornea and the ocular conjunctiva. Imaging can be performed under stable and reproducible conditions and higher quality images are obtained than when using confocal scanning slit microscopy. In vivo "cytology" without staining is made possible by our confocal scanning laser microscope.

Plate 95

a   The Heidelberg Retina Tomograph HRT II as a digital-confocal scanning laser microscope with the external, hydraulically operated Rostock Cornea Module and the video monitoring system (WebCam) for exact positioning of the cornea in front of the contact adapter.

b   Schematic drawing of corneal micromorphology according to Kristic (from: R.V. Krstic, Human Microscopic Anatomy, Springer Verlag 1991, S. 509)

c   Basal cell layer (55 mm from corneal suface) with approximately 11,100 cells per mm²

d   Subepithelial nerve plexus of the cornea with thick, partially parallel and radial cords and fine epithelial nerve fibers after penetration of Bowman's membrane. The distance from thecorneal surface gives exact information about the overall corneal epithelial cell layer thickness (in this example 62 μm)

e   Keratocyte nuclei in the middle corneal stroma approximately 250 mm from Bowman's membrane. Cell density approximately 16 000 cells per mm²

f   The corneal endothelium of the normal cornea with a regular cell density of 2,200 cells per mm²

g, h Epidemic keratokonjunctivitis late stage with intraepithelial lesions are characterized by confocal laser scanning microscopy as intraepithelial space occupying lesions with high reflective nucleus like bodies in the area of wing cell layer (g) and well defind high reflective cellular borders adjacent to Bowman's membrane (h)

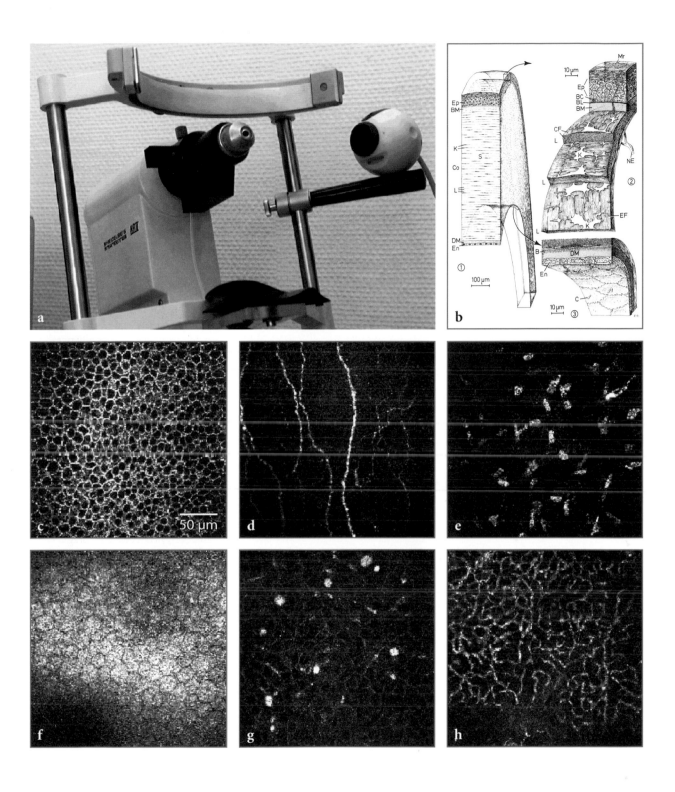

References

Bartsch DU, Intaglietta M, Bille JF, Dreher AW, Gharib M, Freeman WR (1989) Confocal laser tomographic analysis of the retina in eyes with macular hole formation and other focal macular diseases. Am J Ophthalmol 108:277–287

Bathija R, Zangwill L, Berry CC, Sample PA, Weinreb RN (1998) Detection of early glaucomatous structural damage with confocal scanning laser tomography. J Glaucoma 7:121–127

Bowd C, Weinreb RN, Lee B, Emadi A, Zangwill LM (2000) Optic disc topography after medical treatment to reduce intraocular pressure. Am J Ophthalmol 130:280–286

Britton RJ, Drance SM, Schulzer M et al. (1987) The area of the neuroretinal rim of the optic nerve in normal eyes. Am J Ophthalmol 103:497–505

Burk ROW (2001) Laser Scanning Tomographie: Interpretation of the HRT II printout. Z prakt Augenheilkd 22:183–190

Burk ROW, Vihanninjoki K, Bartke T, Tuulonen A, Airaksinen PJ, Völcker HE, König J (2000) Development of the standard reference plane for the Heidelberg Retina Tomograph (HRT). Graefes Arch Clin Exp Ophthalmol 238:375–384

Burk ROW, Noack H, Rohrschneider K, Völcker HE (1998) Prediction of glaucomatous visual field defects by reference plane independent three-dimensional optic nerve head parameters. In: Wall M, Wild J (eds) Perimetry update 1998/1999. Proceedings of the XIIIth. International Perimetric Society Meeting Gardone Riviera, Italy, September 6–9 1998. Kugler, Amsterdam, pp 463–474

Burk ROW, Rohrschneider K, Voelcker HE (1990) Posterior segment laser scanning tomography: contour line modulation in optic disc analysis. Proc SPIE 1357:228–235

Caprioli J, Park HJ, Ugurlu S, Hoffman D (1998) Slope of the peripapillary nerve fibre layer surface in glaucoma. Invest Ophthalmol Vis Sci 39:2321–2328

Chauhan BC (1996) Analysis of changes in the optic nerve head. In: Anderson DR, Drance SM (eds): Encounters in glaucoma research 3. How to ascertain progression and outcome. Kugler, Amsterdam, pp 195–208

Chauhan BC, McCormick TA, Nicolela MT, LeBlanc RP (2001) Optic disc and visual field changes in a prospective longitudinal study of patients with glaucoma. Arch Ophthalmol 119:1492–1499

Chauhan BC, Blanchard JW, Hamilton DC, LeBlanc RP (2000) Technique for detecting serial topographic changes in the optic disc and peripapillary retina using scanning laser tomography. Invest Ophthalmol Vis Sci 41:775–782

Dannheim F, Pelka S, Sampaolesi JR (1995) Reproducibility of optic disk measurements with the Heidelberg Retina Tomograph. In: Mills RP, Wall M (eds) Perimetry update 1994/1995. Kugler, Amsterdam, pp 343–350

Garway-Heath DF, Poinoosawmy D, Wollstein G, Viswanathan A, Kamal D, Fontana L, Hitchings RA (1999) Inter- and intraobserver variation in the analysis of optic disc images: comparison of the Heidelberg retina tomograph and computer assisted planimetry. Br J Ophthalmol 83:664–669

Göbel W, Lieb WE (1997) Quantitative and objective follow-up of papilledema with the Heidelberg Retina Tomograph. Ophthalmologe 94: 673–677

Hatch WV, Flanagan JG, Williams LDE, Buys YM, Farra T, Trope GE (1999) Interobserver agreement of Heidelberg Retina Tomograph parameters as a result of different observers' contour line placement. J Glaucoma 8:232–237

Hudson C, Flanagan JG, Turner GS, McLeod D (1998) Scanning laser tomography z profile signal width as an objective index of macular retinal thickening. Br J Ophthalmol 82:121–130

Hudson C, Charles SJ, Flanagan JG (1997) Objective morphological assessment of macular hole surgery by scanning laser tomography. Br J Ophthalmol 81:107–116

Hudson C, Shah S, Flanagan JG, Brahma A, Ansons A (1995) Scanning laser tomography in benign intracranial hypertension. Lancet 346:1435

Iester M, Mikelberg FS, Coutright P, Burk RO, Caprioli J, Jonas JB, Weinreb RN (2001) Interobserver variability of optic disc variables measured by confocal scanning laser tomography. Am J Ophthalmol 132:57–62

Iester M, Mikelberg FS, Drance SM (1997) The effect of optic disc size on diagnostic precision with the Heidelberg Retina Tomograph. Ophthalmology 104:545–548

Janknecht P, Funk J (1994) Optic nerve head analyzer and Heidelberg Retina Tomograph: Accuracy of topographic measurements in a model eye and in volunteers. Br J Ophthalmol 78:760–768

Jonas JB, Gusek GC, Naumann GO (1998) Optic disc, cup and neuroretinal rim size, configuration and correlations in normal eyes. Invest Ophthalmol Vis Sci 29:1151–1158

Kamal DS, Viswanathan AC, Garway-Heath DF, Hitchings RA, Poinoosawmy D, Bunce C (1999) Detection of optic disc change with the Heidelberg Retina Tomograph before confirmed visual field change in ocular hypertensives converting to early glaucoma. Br J Ophthalmol 83:290–294

Katz J, Tielsch JM, Quigley HA et al (1993) Automated suprathreshold screening for glaucoma: The Baltimore Eye Survey. Invest Ophthalmol Vis Sci 34:3271–3277

Kesen MR, Spaeth GL, Henderer JD, Pereira MLM, Smith AF, Steinmann WC (2002) The Heidelberg Retina Tomograph vs clinical impression in the diagnosis of glaucoma. Am J Ophthalmol 133:613–616

Kotecha A, Siriwardena D, Fitzke FW, Hitchings RA, Khaw PT (2001) Optic disc changes following trabeculotomy: longitudinal and localisation of change. Br J Ophthalmol 85:956–961

Kruse FE, Burk ROW, Völcker HE, Zinser G, Harbarth U (1989) Reproducibility of topographic measurements of the optic nerve head with Laser Tomographic Scanning. Ophthalmology 96:1320–1324

Lesk MR, Spaeth GL, Azuara-Blanco A, Araujo SV, Katz LJ, Terebuh AK, Wilson RP, Moster MR, Schmidt CM (1999) Reversal of optic disc cupping after glaucoma surgery analyzed with a scanning laser tomograph. Ophthalmology 106:1013–1018

Meyer T, Howland HC (2001) How large is the optic disc? Systematic errors in fundus cameras and topographers. Ophthalmic-Physiol-Opt 21:139–150

Miglior S, Albe E, Guareschi M, Rossetti L, Orzalesi N (2002) Intraobserver and and interobserver reproducibility in the evaluation of optic disc stereometric parameters by Heidelberg Retina Tomograph. Ophthalmology 109:1072–1077.

Mikelberg F, Parfitt C, Swindale N, Graham S, Drance S, Gosine R (1995) Ability of the Heidelberg Retina Tomograph to detect early glaucomatous visual field loss. J Glaucoma 4:242–247

Mikelberg FS, Wijsman K, Schulzer M (1993) Reproducibility of topographic parameters obtained with the Heidelberg Retina Tomograph. J Glaucoma 2:101–103

Mulholland DA, Craig JJ, Rankin SJA (1998) Use of scanning laser ophthalmoscopy to monitor papilloedema in idiopathic intracranial hypertension. Br J Ophthalmol 82:1301–1305

Nakla M, Nduaguba C, Rozier M, Hoffman D, Caprioli J (1999) Comparison of imaging techniques to detect glaucomatous optic nerve damage. Invest Ophthalmol Vis Sci 40:397

Orgül S, Cioffi GA, Bacon DR, Van Buskirk EM (1996) Sources of variability of topometric data with a scanning laser ophthalmoscope. Arch Ophthalmol 114:161–164

Rohrschneider K, Burk ROW, Voelcker HE (1994) Follow-up examinations of papillary morphology with laser scanning tomograph. Ophthalmologe 91:811–819

Rohrschneider K, Burk ROW, Kruse FE, Völcker HE (1994) Reproducibility of optic nerve head topography with a new laser tomographic scanning device. Ophthalmology 101:1044–1049

Rohrschneider K, Burk ROW, Völcker HE (1990) Papilledema. Follow-up using laser scanning tomography Fortschr Ophthalmol 87:471–474

Sampaolesi R, Sampaolesi JR (2001) Large optic nerve heads: megalopapilla or megalodiscs. Int Ophthalmol 23:251–257

Scheuerle AF, Specht H, Bueltmann S, Rohrschneider K (2002) Foveal topography and its interaction with ophthalmic wave front analysis. Invest Ophthalmol Vis Sci 43:ARVO abstract 2050

Scheuerle AF, Schmidt E, Burk ROW (2001) Polar contour-line of optic disc border reaches mean retina height (MRH) in normal eyes in laser scanning tomography using the Heidelberg Retina Tomograph (HRT). Invest Ophthalmol Vis Sci 42:135

Scheuerle AF, Steiner HH, Meister C, Burk ROW (1999) Monitoring of papilledema using the Heidelberg-Retina-Tomographen (HRT). Ophthalmologe 96:S120

Schmidt E, Scheuerle AF, Burk ROW, Kruse FE (2003) Analysis of the profile of the peripapillary nerve fiber layer with the Heidelberg Retina Tomograph. Invest Ophthalmol Vis Sci 44:ARVO abstract 5011

Sheen NJ, Aldridge C, Drasdo N, North RV, Morgan JE (2001) The effects of astigmatism and working distance on optic nerve head images using a Heidelberg Retina Tomograoh scanning laser ophthalmoscope. Am J Ophthalmol 131:716–721

Sommer A, Katz J, Quigley HA et al. (1991) Clinically detecable nerve fiber atrophy precedes the onset of glaucomatous field loss. Arch Ophthalmol 109:77–83

Swindale NV, Stjepanovic G, Chin A, Mikelberg FS (2000) Automated analysis of normal and glaucomatous optic nerve head topography images. Invest Ophthalmol Vis Sci 41:1730–1742

Trick GL, Vesti E, Tawansy K, Skarf B, Gartner J (1998) Quantitative evaluation of papilledema in pseudotumor cerebri. Invest Ophthalmol Vis Sci 39:1964–1971

Uchida H, Brigatti L, Caprioli J (1996) Detection of structural damage from glaucoma with confocal laser image analysis. Invest Ophthalmol Vis Sci 37:2393–2401

Weinberger D, Stiebel H, Gaton DD, Priel E, Yassur Y (1996) Three-dimensional measurements of central serous chorioretinopathy using a scanning laser tomograph. Am J Ophthalmol 122:864–869

Weinberger D, Stiebel H, Gaton DD, Priel E, Yassur Y (1995) Three-dimensional measurements of idiopathic macular holes using scanning laser tomography. Ophthalmology 102:1445–1449.

Weinreb RN, Lusky M, Bartsch DU, Morsman D (1993) Effect of repetitive imaging on topographic measurements of the optic nerve head. Arch Ophthalmol 111:636–638

Wollstein G, Garway-Heath DF, Fontana L, Hitchings RA (2000) Identifying early glaucomatous changes. Comparison between expert clinical assessment of optic disc photographs and confocal scanning ophthalmoscopy. Ophthalmology 107:2272–2277

Wollstein G, Garway-Heath DF, Hitchings RA (1998) Identification of early glaucoma cases with the scanning laser ophthalmoscope. Ophthalmology 105:1557–1563

Zinser G, Wijnaendts-van-Resandt RW, Ihrig C (1988) Confocal laser scanning microscopy for ophthalmology. Proc SPIE 1028:127–132

Acknowledgements

We are grateful for the help and support that we have received from the faculty and staff of the Department of Ophthalmology at the University of Heidelberg. We have been extraordinarily lucky in having H.E. Völcker as our chairman. He provided all the means to equip the facility with the up-to-date apparatus that was indispensable for this book`s development. The "Atlas of laser scanning ophthalmoscopy" reflects many of the skills and experience we learned from our great teacher Reinhard O.W. Burk. Without his influence, we never would have started this project.

We are appreciative of the expert work and support rendered by the company Heidelberg Engineering GmbH and are especially indebted to Gerhard Zinser, who was always willing to share his profound knowledge with us.

We would like to thank all of our collegues who helped us to acquire the wealth of material we wanted to present to a broader audience. Also, we wish to express particular thanks to Daniel Miller, who patiently proofread the manuscript.